# Lands Forlorn

# Lands Forlorn

A Story of an Expedition to

Hearne's Coppermine River

By

George M. Douglas

with

Douglas's journal and maps from his

First Journey to the Coppermine

and

Some letters pertaining to the expedition

Edited by

Robert S. Hildebrand

Zancudo Press
Tucson, AZ
2008

Published in the United States by Zancudo Press, 1401 N. Camino de Juan, Tucson, AZ 85745

Front cover painting and design. Gregory K. Crawford

LIBRARY OF CONGRESS CATALOGING-IN-PUBLICATION DATA

Douglas, George Mellis
Lands Forlorn: the story of an expedition to Hearne's Coppermine river/ by George M. Douglas: edited by Robert S. Hildebrand
448p.   ISBN 978-0-615-19529-2

1. Exploration—Northwest Territories—Coppermine River Valley—History.
2. Geology—Basalts—Coppermine River Lavas—Coppermine River Valley. 3. Mining—Copper—Basaltic Lavas—Northwest Territories—James Douglas. 4. Northwest—Canada—Description and travel. 5. Travelers—Northwest,—Canadian—Biography. 6. Adventure and adventurers—Northwest—Canadian—Biography. l. Title.


Lands Forlorn web site: www.landsforlorn.com

Lionel and George Douglas, 1911

# Foreword to the Zancudo Press Edition

## by Robert S. Hildebrand

*Lands Forlorn* is an exceptional book written by an extraordinary man. Although nearly a century has passed since its publication, it remains one of the most lucid and compelling tales of northern exploration ever written – a testament to the prowess of its author, George Mellis Douglas.

The book Douglas first had in mind was to be a straightforward narrative of a small and focused prospecting trip to the Coppermine Mountains of far northern Canada, but later events arising from the trip, such as murder and starvation, animated *Lands Forlorn* with both drama and historical value. Yet the book is not sensationalist, for it does not emphasize suffering, hardship, and the lonely deaths that made accounts of even the most trivial of explorations by incompetent explorers well–known. Instead, the book is whimsically understated, describing in delightful detail the personalities, routines, and discoveries of three men on their first journey north. In that way *Lands Forlorn* reflects the necessity of a pragmatic outlook on such journeys, for those who have worked or traveled in the northland will attest that even with adequate foresight and planning, survival and achievement north of the Arctic circle are adventure enough in themselves. Thus, Douglas never brags about how difficult things were, nor on the other hand trivializes them, but instead presents an honest depiction of events, interlaced with a healthy dose of humor and astute cynicism.

George Douglas was born in Halifax, Nova Scotia in 1875, the first child of Campbell Mellis Douglas and Eleanor Burmester. His brother, Lionel Dale Douglas, was born three years later, in 1878. Burmeister, niece of Sir Edward Belcher, scientist, explorer and author, who later in life was put in charge of the Franklin search, had two children, Bryce and May McMaster, by a previous marriage. Her first husband, Valentine McMaster, was awarded the Victoria Cross for actions in India but died of typhus in 1859. Douglas and Burmeister had two more children, Muriel and Evan, before Eleanor died in 1894.

The Douglas family came to Canada from the United States and Scotland during, and shortly after, the American Revolution. George's father,

Campbell, and grandfather, also George Mellis Douglas, were both famous physicians. For 25 years his grandfather was the doctor in charge of the Canadian Grosse Ile Quarantine Station where thousands of immigrants arrived, many carrying typhus and cholera. His father, an expert with small boats, was awarded the Victoria Cross for rowing through extreme seas time and time again in a gig – a skiff–like boat – to rescue 17 men who were stranded ashore and threatened by natives on Little Andaman Island during the Sepoy Rebellion. In 1885, while serving as a doctor during the North–West Rebellion he used a folding canoe of his own design to save wounded soldiers trapped on the steamer, *Northcote*, which had run aground in the Saskatchewan River. Later, he took the folding canoe to England and used it to cross the English Channel. The family was clearly of superb stock, and the sons must have felt some pressure to live up to the family name.

When the boys were young, the Douglas family moved often, but in 1883 when George was eight, the family settled at Northcote Farm on the shores of Katchiwano Lake [now Katchawanooka], Ontario. It was an idyllic place for George and his younger brother, Lionel, to live, as they could explore the bays, islands and nearby forests. Northcote was a real farm in those days with cows, sheep, horses, wagons and sleighs, so there were chores as well as schooling. Their mother, despite the demands of isolated farm life, was a wonderful influence, encouraging the children to read, play musical instruments, and foster friendships.

It was here at Northcote, that George and his brother learned canoeing, camping, sailing, hunting, and fishing – disciplines that would serve them well on their northern adventure. In those days only a dirt road and an old Indian trail connected Northcote with the outside world, so travel to school or the store was by foot, boat, skis or snowshoes. In 1887, when George was twelve, he started riding a homemade bicycle – a strange old beast made by his father with solid rubber tires and a saddle set on a long spring – and he was still cycling well into his sixties.

George attended Sheldrake's School [now Lakefield College School] in nearby Lakefield; spent time with his wife's mother in

Halifax studying for naval examinations, which he ultimately failed; and in the fall of 1890, went to Trinity College School in Port Hope.

In 1894 George's mother died and his father, whose folding boat business had failed, sold Northcote and took the boys to England, where he rejoined the army as a doctor. George, who had started engineering classes at The University of Toronto the previous fall, was crushed at

Farm hand, George, Lionel, and Campbell Douglas at Northcote, 1888

the sale of Northcote and "disgusted at my father's decision to move back to England and pull me especially up by the roots!" He was terribly homesick and after living in Canada for so long it must indeed have been a rude awakening. He consistently sought relief from the industrialized bleakness of northern England in Thoreau's books and diaries, with *The Maine Woods*, his favorite. Whenever he could, he escaped by bicycle, often taking rides of 100 or more miles, sometimes starting out at 2 A. M.

Moving also must have been a shock to Lionel, for upon arrival in England he immediately signed on to the merchant training ship *Conway* for two years of marine schooling and then, in 1897, went to sea on the four-masted barque *Silberhorn*. Before 1901, Lionel had experienced three voyages round Cape Horn to the west coast of North America and back.

While Lionel was training for a life at sea, George attended Rutherford Technical College in Newcastle-on-Tyne. Afterwards, he served a three-year apprenticeship with the engineering firms of Amstrong, Whitworth & Co. and Hawthorn, Leslie & Co., both of Newcastle-on-Tyne. There, he worked mostly on warship engines for sea trials. For the first time George was plunged into the thick of modern industrialism in what he considered "its most sordid surroundings, in the north of England, and in its most abhorrent form, the making of armaments solely for profit". He so detested working on anything to do with the implements of war that he quit his job and took up marine engineering, serving three years at sea, first with the Allan Line and then on two trips around the world in a year with the White Star Line.

In 1901, having had all he could take of both sea life and England, George found employment in a remote area of Sonora, Mexico, as chief engineer of the Montezuma Copper Company. There he worked for his cousin, James Douglas, head of the company and famed Arizona copper baron. George was in charge of a gas plant and central power house utilizing large gas engines.

Cousin James, as he was known to the Canadian Douglas family, was a remarkable man of great intelligence and ambition. He started out as a Quebec preacher, but ended up a copper baron, head of Phelps-Dodge Corporation, and Chancellor of Queen's University. He had money, influence, and a genuine love for his Canadian branch of the family. He was always available to help in any way he could and often did so.

While stationed in Mexico, George also traveled widely, ranging from New England to Lower California, and from Mexico to Milwaukee. For him, Northcote was always home and he was continually drawn back there, visiting whenever he had the chance, and taking many photos, which today serve to document how much he missed the place of his upbringing.

He continued working in the Southwest and spent 1904 inspecting the construction of large gas engines for Detroit Copper Mining of Morenci, Arizona, before heading to Wisconsin to supervise construction for the Power & Mining Machinery Co. of Wisconsin. He returned to Arizona in 1906 as a consulting engineer and was sent on a tour of Europe to

Northcote in 1909

research diesel engines. While in Europe he spent time resting at his father's place in Essex. Nevertheless, his incessant traveling, coupled with the tragic loss of his Mexican fiancée to tuberculosis, destroyed his health and he was saved only by a retreat to Northcote, which he had bought on his way to England in 1907. Northcote remained "home base" until just before his death in 1963.

While George was moving back and forth between Canada, Europe, and the Southwest, Lionel remained at sea, working as a mate on various Clan Line freighters. In April of 1905 he signed on with Canadian Pacific Railway's *SS Montezuma*, which turned out to be the beginning of a long and successful career under CPR's red and white checkered house flag. For Lionel, "home base" became Vancouver, British Columbia. In July 1905 he was transferred to the *Empress of India* as 3rd Mate, and as the years passed, Lionel would continue to climb the command ladder for CP, serving as mate on the *Empress of China* and the *Empress of Japan*, before becoming Chief Officer of the *Empress of Japan* in 1910. It was then that he was introduced to Christine Peirce of Bangor, Maine, who was embarking on a tour of China, Japan, India and Southeast Asian. Smitten by her beauty, Lionel pushed for a romantic relationship, but she was reluctant to take up with a young sailor.

From 1909 to 1911 George worked winters for Phelps, Dodge & Co. in Arizona and Mexico as a consulting engineer on all manner of engines, compressors, and gas plants. He spent the summers of 1908–1910 at Northcote, and was excited to be planning a canoe trip along an old voyageurs' route with his father for the summer of 1910. Suddenly in December 1909 he received the shocking and surprising news from his Cousin James that his father was near death. He rushed to New York by train only to learn that his father had died. It was while in New York for this unhappy duty that Cousin James, then president of Phelps Dodge Corporation, offered to grubstake the grief–stricken George if he explored and prospected the Coppermine Mountains.

The Coppermine Mountains are a low chain of hills really, that strike more or less east–west just south of the Arctic coast near the mouth of the Coppermine River. The area was at that time, and still is, one of the more remote places on the continent. *Lands Forlorn* is the story of how George Douglas, along with his brother Lionel and a geologist, August Sandberg, as they traveled to the area, searched for copper, and returned to civilization after successfully "wintering over" north of the Arctic circle.

First, they used stagecoaches, scows, and steamers of the Hudson's Bay Company to travel from Edmonton to Fort Norman on the Mackenzie River. Douglas's descriptions and photographs of this commercial travel are vivid and delightful, and together with Agnes Dean Cameron's *The New North*, form a nearly complete picture of commercial water travel in the Northwest Territories at the turn of the 20th century.

At Fort Norman, where the three explorers left commercial transportation to strike off on their own, they met Father Jean–Baptiste Rouvier, an Oblate priest sent to convert the Inuit in the Coppermine River area, and John Hornby,

Geologist August Sandberg and George Douglas in Mexico, early 1900's

an Englishman, who had just arrived from Great Bear Lake to assist and guide Rouvier. The Douglas group hired some Indians and tracked their York boat and gear up the swift–flowing Great Bear River to Great Bear Lake. They sailed across the lake – enduring an intense storm that drove them wildly off course, passing ice floes in thick fog, and nearly running aground – to finally arrive at the northeasterly corner of Dease Arm at the mouth of the Dease River, which they sailed up until they ran aground. There, Lionel and an Indian assistant built a sturdy cabin, while George and Sandberg canoed, waded, and portaged up the Dease River and its tributaries to the Dismal Lakes. They then descended the Kendall River to the Coppermine where they explored the Coppermine Mountains for a week before returning to their newly–constructed cabin.

John Hornby and Father Rouvier arrived in the area just after the Douglas party and built an inferior cabin nearby. The two parties visited often during the winter, and the following spring Hornby traveled with the Douglas group by dogsled and on foot to the Coppermine Mountains. The Douglas brothers continued alone on foot to the coast. After this successful trip the group then returned to the Dease and after waiting weeks for the ice to break up crossed Great Bear Lake in a canoe to arrive at Fort Norman in August of 1912.

Although it was the trip of a lifetime, and George fully expected to return to the Coppermine Mountains for additional prospecting, it was not to be, for the area was soon closed to mineral development by the Canadian government. Nevertheless, the brothers needed to work, so within a month of arriving at Northcote, Lionel returned to Canadian Pacific, where he was posted aboard the *Empress of Japan*, and George headed back to southern Arizona to work for Phelps–Dodge. The following spring found George back at Northcote where he bought a small island (Wee Island) in Stoney Lake.

While at Northcote, George began work on *Lands Forlorn*, and although he found it the most difficult thing he had ever done, was able to complete a draft by the following April. The book was published in July, just days after the assassination of Archduke Franz Ferdinand in Sarajevo. That month Austria–Hungary declared war on Serbia and the Great War had begun. Hornby soon arrived at Northcote on his way to enlist, and George, instead of heading north again as he had expected, traveled to New York where he tried to volunteer for service at the office of the British Consulate General. He was devastated when he was rejected for his slight deafness; after all, he was a better shot and had more experience outdoors and with a variety of engines and boats than nearly anyone else who could have been selected.

In spite of this setback to his ego, George was stubborn, so he traveled to England, where he hoped to join the Royal Flying Corps as a pilot, or possibly to serve as an engineer in the Royal Navy; but once again, and to his further disappointment, he was rejected because of his deafness. Instead, George was appointed to one of Amstrong's armament factories, which, given his reaction when working there before, must have caused him considerable consternation. Nevertheless, his dismay was short–lived because he resigned shortly after arriving there when a fellow employee was shabbily treated by management.

In early 1915 he spent six months opening mines around Tombstone, Arizona before returning to Northcote for a year. September of 1916 saw him back in Mexico as a consultant, but he returned to Canada the following April to marry Frances Mackenzie [also known as Kay], who had spent summers near Northcote with her family. The newlyweds promptly headed to Mexico for the winter.

A few months after arriving in Mexico, they learned that August Sandberg had been jailed in Arizona as a German sympathizer because he was too headstrong to stop touting the superiority of German engineering. George and Frances rushed to Arizona and managed to get Sandberg released, but by that time his job was given to another. He was so distraught by the unfairness of it all that he abandoned his life in Arizona and moved to the Imperial Valley of California to live with his brother. What then became of Sandberg is unknown.

While George continued his pre–trip pattern of wintering in the Southwest and spending summers at Northcote, Lionel advanced through the ranks of Canadian Pacific. It didn't take too long after returning from the Arctic before Lionel was named Chief Officer of the new CPR steamship Empress of Asia. Apparently the promotion impressed Christine – and presumably her parents – for they were married in March of 1913. Lionel didn't see combat because he was traveling around the world transporting men and materiel. During this period shore leave was productive as he and Christine had two sons. After the war Lionel was promoted to Commander, and commanded steamships plying the Pacific until 1940, when he retired.

At Northcote, George Douglas received information about some of the characters he had met on his Coppermine trip. First, he learned that Harry Radford and his companion Tom Street had been killed. They left Fort Smith on June 27, 1911, not long after meeting the Douglas party there. The following year they were along the Arctic Coast when Radford, who was known to have trouble with natives, misunderstood when a native bearer dropped out because his wife broke a leg. Radford beat the man with a whip. Infuriated, members of the tribe speared Radford to death and killed Street as he ran for his rifle. The Mounties traveled more than 5000 miles over two years to investigate the possible crime, and once they learned what had happened, declined to arrest the perpetrators.

Douglas also received a long letter from RCMP corporal Denny LaNauze, who had traveled south by steamer with the Douglas party in 1912. In his letter LaNauze described his expedition to investigate the killings of Father Rouvier and his new partner, Father LeRoux, near Bloody Falls in the spring of 1913. LaNauze was the Mountie in charge of the investigation and he and his large party traveled up the Great Bear River, across Great Bear Lake in a York boat, and followed more or less the same path to Coronation Gulf as the Douglas party. They soon found the two Inuit who were responsible and brought them to Edmonton for trial.

In 1928 George carried out a summer journey by canoe along the southeastern shore of Great Slave Lake for United Verde Copper Company, an Arizona mining company owned by James "Rawhide Jimmy" Douglas, one of Cousin James's sons. Douglas and his prospecting partner, Carl Lausen, explored the east arm of Great Slave Lake, but had no luck in locating mineral deposits. Their trip almost ended in disaster when a forest fire swept through their unattended camp. They spotted the column of smoke from afar and rushed back to camp, arriving just in time to keep the fire from reaching their other boat and their gasoline cache, but too late to save their tent, bedding, clothes, and other important necessities. Stormy weather forced them to spend the next nine days in spruce lean-tos on a nearby island before they could leave the area. Douglas, in typical understated fashion, described the event in a telegram to his wife: "lost entire camp by fire, fortunately had canoes, food and gas, weather turned wet and stormy homeward".

That fall, when Douglas returned home, he learned that John Hornby, his young nephew, and another boy, had died east of Great Slave Lake. They tried to winter over there but there was no game, so they all starved. The death of

Hornby deeply disturbed George Douglas, for even though he knew that the eccentric, headstrong Hornby always lived on the edge, he was nevertheless emotionally attached to him and considered him a sort of wayward son.

For the rest of Douglas's life, nearly everyone he met wanted to know about Hornby, yet Douglas knew too much, and so generally declined to talk about him. Finally, in 1955 after hearing a radio program that glorified Hornby's abilities, he became irritated enough to contact the author and set the record straight. In the end, he contributed at least half the material for George Whalley's now classic book, *The Legend of John Hornby*.

After his trip in 1928, Douglas did not return north again until 1932 when he made two trips to stake claims on coal beds that he had discovered in 1912 on the north shore of Great Bear Lake. First, there was a rapid stealth trip by air during the winter with famed bush pilot Walter Gilbert. They stopped at LaBine Point for a night, and in this boom period of the exploration and development of the recently–discovered uranium–silver deposits, Douglas felt welcome and at home among the misfits, prospectors, miners and bush pilots with the same spirit of camaraderie that he had felt in the early days in the Arizona copper district.

The following summer, Douglas and his assistant Peter Pitcher, a university student who later went on to manage the Giant–Yellowknife gold mine, headed up the Bear River in an outboard–powered canoe, which as usual was especially designed and built for Douglas. They first headed for Jupiter Bay where Douglas had found float coal when stormbound there in 1911. Finding no coal beds, they returned to Franklin and turned northward along the west shore of Keith Arm. They spent a few weeks at Douglas Bay near Etacho Point (Gros Cap) surveying the claims Douglas had staked the previous winter and then returned to Norman. Although it was only a matter of a few days' motoring to Hodgson's Point, and Douglas would have liked to revisit the spot, he understood with a measure of sadness that he could never recreate the feelings he had once experienced there.

In 1935, Douglas again returned north, this time to the north shore of Lake Athabasca and the southeastern shore of Great Slave Lake as part of another summer's prospecting. His assistant was Bobby Jones. After completing their explorations in the east arm, they camped on an island about 10 miles northeast of the mouth of the Taltson River and just offshore from the impressive MacDonald fault scarp. In 1928, when he and Lausen camped there, Douglas had named it Eagle Island. They were waiting for a plane to transfer Bobby south to join another field party. Douglas himself was to wait for yet another plane to transport him and the incomparable northern surveyor, Guy Blanchet, north to Great Bear Lake to reduce the acreage of the coal claims he had staked and surveyed in 1932. The renowned bush pilot Wop May flew in late one evening, picked up Jones, dropped off mail for Douglas, and quickly headed south, leaving Douglas alone on the island with a handful of mail. To Douglas there was nothing better than to be left alone on an island in the middle of nowhere and he gleefully wrote, "This sudden interruption and sudden disappearance of 'civilization' into my detachment from the world of men was a peculiar experience. I cannot say that I felt at all lonely when the plane took off, but rather a certain relief, and perhaps even a decided satisfaction to be myself: *Nunquam minus solus quam cum solus.*"

His tranquility was short-lived, however, as within a day pilot Matt Berry swooped down from the sky. With him were Blanchet and prospector Jack Stark, a strange fellow who wore a suit even while prospecting in the bush. He was to be dropped at Fort Rae on their way north. After a night at Fort Rae due to bad flying weather at their destination, they flew to the small mining town of Cameron Bay, located a few short miles east of the radium–silver mine at LaBine Point. Within minutes of arriving at Cameron Bay another plane — this one piloted by famed pilot Punch Dickins — arrived carrying Dr. Charles Camsell. Douglas had known Camsell since the 1911–12 trip when he sought out Camsell's advice because Camsell had made a trip to the same area in 1900 with a Geological Survey of Canada field party. They all agreed that it was a notable day for Cameron Bay when three real old–timers of the north, along with such pioneers of northern aviation as Punch Dickins and Matt Berry met there.

After a week–long trip to Douglas Bay, Blanchet and Douglas returned to Eagle Island in Great Slave Lake, where Douglas was "grateful to be back again to our well found camp and to find it well, content with canoe travel and camping, and above all, glad to be done with planes, and once more master of my own movements".

Douglas's last trip north was in 1938, when he took standard Hudson Bay Co. water transport to Fort Smith, and then traveled by canoe along the southeast shore of Great Slave Lake to the Chuban River, where he and René Hansen looked for mineralization. Having no luck there they returned to Great Slave Lake and cruised through the islands to the north shore, then on to the new, raw townsite of Yellowknife, where all hell had broken loose following the recent discovery of gold. They camped on a nearby island and were joined by another prospector named Tom Greenland.

After a few days in Yellowknife watching more float planes land and take off than they had ever seen in one place, the group quietly canoed up the north arm of Great Slave Lake to Fort Rae and into Snare River country, where they prospected for a few weeks. On their return, Douglas, who likely realized that this was his last northern trip, left the group at Yellowknife and took a pleasure cruise aboard the H.B. Co. *Dease Lake* to Resolution and Reliance at the eastern end of the big lake.

When Douglas returned to Yellowknife he was flabbergasted at the results of the gold frenzy and called it "the most visibly drunk place I have ever seen". Given the combination of great riches and bad behavior he witnessed that summer, he obviously had mixed emotions about exploration and mining when he wrote in his final report, "I regret to say that our season's work must be regarded as barren, and I am not even prepared to make any recommendation as a result of it."

For Douglas what mattered most was not the results, but the journey, which had to be done in proper style and form. In *The Legend of John Hornby*, George Whalley wrote that, "In certain matters, George Douglas was extremely fastidious: he had a keen sense of occasion and an exquisite sense of order". There is no question that Douglas was a meticulous, at times even arrogant, man who didn't suffer incompetents or the unprepared well; but he was always kind and generous, even to those he didn't understand. Neither John Hornby nor Prentice Downes - whom he met on this trip- were ever well–prepared, nor were they neat in any sense of the word, yet Douglas had firm friendships with both and a long relationship with Downes. He understood that both men had real substance and found the North as alluring as he did.

Almost certainly, Douglas understood from an early age that most people were more interested in appearances than substance, but perhaps it was never driven home to him more than when his group arrived in Fort Norman after their trip and were basically ignored. He summed up the style–over–substance reality of life when he stated with a healthy dose of cynicism, "It is true that we had failed to conform to the the convention requiring that the explorer should come to the first post ragged and half–starved, eating his moccasins and mitts. So it was really all our fault in both cases, and we only got the proper punishment that in some shape or other is invariably meted out to all offenders against convention."

He was clearly a romantic, far more interested in getting to the shores of the Arctic Ocean than he was in getting wealthy by prospecting. He had an inflexibility that arose directly from his incorruptible integrity, which made others realize that he could be a trusted friend and loyal confidant. He was extremely generous, both with his time and property, and despite his size was a very gentle man. He was never loud, but had a firmness that could only have come from sparring with his father.

Because Douglas had such strong personal integrity, he was completely unable to compromise, which naturally caused him trouble in his professional life and resulted in a self–imposed solitude in his private life, for small talk offended all his sensibilities. As he wrote to Downes, "On what common grounds can I, who walk 3 miles to get some potatoes, who reads the best English papers (or Thoreau!), meet people who take the latest model car if they have to go half a block, whose pampered life is supported by war economy, whose idea of social entertainment is a cocktail party of 30 or more in space inadequate for 6 and the air filled with cigarette smoke — what the devil would I do in that galley?"

Douglas used forays into wilderness to attain what he considered insight and wisdom. He preferred to live with nature — often moving with the seasons,

George Douglas at Northcote, late 1950's

and always with a keen awareness and appreciation of his surroundings- for it was in the outdoors that he felt truly whole. His letters and journals are filled with observations about weather, birds, rocks, and ice; yet these were not recorded in dry fashion, but rather presented with a keen sense of interest and desire to understand natural patterns and phenomena. It was this philosophy, and the comfort he found in natural solitude, that in many ways governed his later years at Northcote, for he and Kay lived a life without electricity, appliances, or running water in a series of cabins and houseboats depending partly on the season, partly on the weather, and always on whim. They moved from Winter Quarters to Tool House, a log cabin built in the 1840's, to their houseboat, *The Floater*, to Sing Kettle Island or Wee Island. They read, gardened, wrote letters, took photographs,

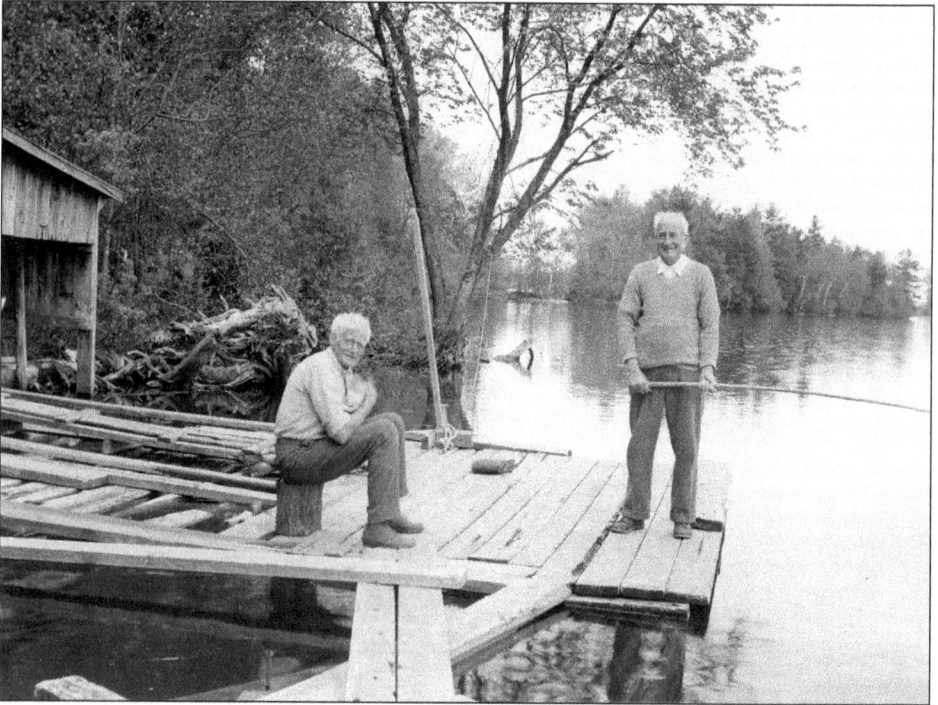

George and Lionel on the boathouse dock, 1950's

played chess by mail, played violin, cared for their boats and patched their own clothes — simply enjoying life on their terms with as few intrusions from the modern world as possible. Douglas understood how strange their life was when he wrote, "It is hard for normal people to realize the primitive way we live. Even Rawhide Douglas, who stayed with us occasionally couldn't grasp it and used to make suggestions as to what we ought to do that were impossible for us – or for me. It requires years of training and knowledge to be a guest at Northcote* and people especially need to give us several days notice of proposed visits." Nevertheless, they always welcomed visitors.

In fact, George was rather social, but entirely on his own terms. He loved his correspondence with northern friends and acquaintances. He was always available to provide advice and often did so, whether it be to Queen's English professor George Whalley as he researched his book on Hornby, or Ann Marie Krougher, a geography student at McGill traveling to Great Bear Lake for anthropological research. His letters, and mostly unpublished writings over the years, now provide a wealth of information on a broad range of topics.

Early in 1962, Lionel Douglas passed away in Vancouver and on June 6, 1963 George Douglas died in his sleep. Both brothers lived exceptional lives, yet remain little known outside of canoeing circles and among a few northern history enthusiasts. I hope this book helps to give them renewed recognition.

*As this book goes to press, Northcote Farm, with its 210 acres and 2000 feet of lakefront, was purchased and donated to Lakefield College School, George's *alma mater* in order to preserve it from development.

# About this edition:

In the early 1990's, after 20 years of field work in the Great Bear Lake–Coppermine River area, I started to research and write a monograph on the complete history of that country. It is a truly remarkable area and has documented history that ranges from about four billion years ago to the present. It was during this research that I met the remarkable Frances Douglas and she assigned me the rights to *Lands Forlorn*. The publication of this volume has thus been a fifteen–year labor of love, for the number of photographs was sufficient to cause most publishers to see it only as a money–losing project. Therefore, I started to scan all the photographs and prepare the text with the idea of putting it on the web or self–publishing it on demand. Finally, we decided to use our own publishing company, Zancudo Press.

In this new edition, I tried to do justice to the original book. First, I kept the original format because it was well done to begin with, and given the number of photographs, I didn't feel that it could be significantly improved upon. In this digital age exact matches for old lead type simply do not exist, so I settled on a close substitute of a font that at least existed at the time, Century Old Style. With assistance from the expert community at typophile. com I managed to identify and track down a digital version of the old Fraktur font used for the page headers.

The photographs were a challenge. Wherever possible, prints were made directly from duplicates of the cellulose nitrate negatives developed by Douglas and currently stored in the National Archives of Canada. Where these were not available I worked from copy negatives of existing prints taken from family photo albums generously provided by Wayne Dutcher, grandson–in–law of Lionel; for the few photographs for which I could not locate prints or negatives, I scanned one of my original printed copies of the book with obviously inferior results. Most of the original negatives were filthy and had deteriorated in the 80 or so years since they were developed so the amount of digital restoration and cleaning was significant.

On their spring trip to the Coppermine, Douglas's camera developed a light leak that caused a flare in the lower left–hand quadrant of the photographs. This caused variable damage to the images depending on the amount of bellows extension and the exposure time. It is hard to imagine that Douglas didn't notice this when he developed the photographs at the cabin on Hodgson's Point just prior to traveling across Great Bear Lake; but strangely he didn't fix the leak by putting some electrical tape, which he had (p. 189), over the hole in the bellows. Possibly he had used all the tape by this time or possibly he had left it in one of the caches on the Coppermine. Whatever the reason, the same problem exists on photographs taken during the return journey to civilization. For the original 1914 edition, many of these, but not all, were retouched by an artist. Because the original negatives still had the original flaws, I attempted to restore them by various means so as to make them not so obvious to the modern reader.

This edition also contains previously unpublished material. First, there are selected letters from the principals and the sponsor of the expedition, James Douglas. They are included not only for the sake of completeness, but also because they provide a much better understanding of the planning and preparatory stages of the journey. Likewise, I included Douglas' diary from his first trip to the Coppermine. It makes for fascinating reading, has wonderfully-detailed maps drawn by Douglas, and is markedly different from the account published in the book. No doubt you'll be struck, as I was, by the terrible weather they had on this journey.

In a few places I added punctuation to improve the readability of the diary, but otherwise the text is presented as typed by Douglas. It is important to recognize that these are journals and letters written in boats and tents, not finished text.

If I had any complaint about the original book it would be the lack of detailed maps showing camping spots, routes of exploratory hikes, and so forth. While I have detailed maps and air photos, most don't, so the inclusion of Douglas's superb maps from the journal, drafted in pen and ink and colored with watercolors, but here reproduced in black and white, helps remedy that shortcoming. Color scans of the originals can be found at landsforlorn.com

# Acknowledgements:

First and foremost, I would like to thank Dr. Paul F. Hoffman for introducing me to Canada's northland. I owe a lot to John C. McGlynn, lamentably now deceased, and Bill Padgham, both of whom supported my own field research in the northland for many years.

I thank Mrs. Frances Douglas, for not only was she an early source of information on her husband, but this volume is a direct result of her coaxing. Moreover, we all owe her thanks for the orderly fashion in which she sorted and allocated papers and photographs to various public repositories.

Enid Mallory made arrangements for Bob Baragar and me to first visit Northcote in 1993. Kathy Hooke, the unofficial keeper of all things Douglas, generously researched old letters and provided photographs. Wayne Dutcher, grandson–in–law of Lionel, provided photographic prints and a diary from the journey, as well as a biographical sketch of Lionel Douglas that greatly improved my understanding of his life.

Michael Peake, Monika Scheube, and Tom Frisch all contributed in smaller, but no less important, ways. Three wonderful authors, Robert Cockburn, McKay Jenkins and Wayne Fields, read the foreword. I must thank Sam Bowring, not only for his constant encouragement, but for sharing countless hours in the arctic twilight of Canada's northland.

And finally, I want to thank my wife, Emirse, who learned new software and built the book. The project is much the better for her dedication and tireless attention to detail.

# Photography Credits

# Credits

190380; 30: GMD NAC PA 190379; 31a: GMD NAC PA 190398; 31b: GMD NAC PA 190397; 32: GMD NAC PA 190426; 33: GMD NAC PA 190425; 34: GMD NAC PA 190377; 35: GMD NAC PA 190376; 36: GMD NAC PA 190375; 37: GMD NAC PA 190374; 38: GMD NAC PA 190349; 39: GMD NAC PA 190427; 40: GMD NAC PA 190396; 41: WD; 42: GMD NAC PA 190424; 43: GMD NAC PA 190395; 44: GMD NAC PA 190417; 45: GMD NAC PA 145191; 46: GMD NAC PA 145190; 49: GMD NAC PA 190373; 50: WD; 51: WD; 52: GMD NAC PA 190348; 53: GMD NAC PA 190347; 54: GMD NAC PA 145213; 55: GMD NAC PA 145212; 56: GMD NAC PA 190372; 57: GMD NAC PA 190371; 58: GMD NAC PA 190394; 59: GMD NAC PA 145211; 60: WD; 61: GMD NAC PA 190370; 62: GMD NAC PA 120585 ; 63: GMD NAC PA 190369; 64: GMD NAC PA 190368; 65: GMD NAC PA 145210; 66: GMD NAC PA 190367; 67: GMD NAC PA 190366; 68: GMD NAC PA 190421; 69: GMD NAC PA 190365; 70: GMD NAC PA 190346; 71: GMD NAC PA 190364; 72: GMD NAC PA 190363; 73: GMD NAC PA 145204; 76: GMD NAC PA 190393; 79: GMD NAC PA 145209; 84: GMD NAC PA 190392; 85: GMD NAC PA 145208; 86: GMD NAC PA 190362; 87: GMD NAC PA 190361; 88: GMD NAC PA 190391; 89: GMD NAC PA 120595; 91: GMD NAC PA 145207; 92: GMD NAC PA 190360; 94: GMD NAC PA 190359; 95: GMD NAC PA 190358; 96: GMD NAC PA 190390; 98: GMD NAC PA 145221; 99: GMD NAC PA 189785; 100: GMD NAC PA 187355; 101: GMD NAC PA 190357; 105: GMD NAC PA 145219; 107: GMD NAC PA 120586; 108: GMD NAC PA 120587; 111: GMD NAC PA 145206; 112: GMD NAC PA 145220; 115: GMD NAC PA 190423; 116: GMD NAC PA 190345; 117: GMD NAC PA 145222; 118: GMD NAC PA 120588; 119: GMD NAC PA 190344; 120: GMD NAC PA 190356; 121: GMD NAC PA 189773; 122: GMD NAC PA 190355; 123: GMD NAC PA 189772; 124: GMD NAC PA 189771; 125: GMD NAC PA 150888; 129: GMD NAC PA 189770; 130: GMD NAC PA 190354; 134: GMD NAC PA 142561; 135: GMD NAC PA 190353; 141: WD; 142: (LF); 144: WD; 145: GMD NAC PA 150891; 146: (LF); 152: GMD NAC PA 150881; 153: WD; 154: GMD NAC PA 187564; 155: GMD NAC PA 187354; 161: (LF); 163: GMD NAC PA 120592; 164: GMD NAC PA 190389; 165: WD; 167: GMD NAC PA 147451; 168: GMD NAC PA 120589; 171: WD; 172: GMD NAC PA 150874; 173: GMD NAC PA 187565; 174: WD; 177: WD; 180: WD; 181: WD; 183: GMD NAC PA 150890; 186: GMD NAC PA 190407; 187: GMD NAC PA 189763; 188: GMD NAC PA 190388; 190: WD; 191: GMD NAC PA 190387; 192: GMD NAC PA 189762; 193: WD; 196: GMD NAC PA 190408; 197: GMD NAC PA 150892; 198: GMD NAC PA 190206; 199: GMD NAC PA 189768; 200: GMD NAC PA 190405; 201: GMD NAC PA 189767; 202: GMD NAC PA 150909; 203: GMD NAC PA 120590; 204: GMD NAC PA 190404; 205: (LF); 207: (LF); 208: GMD NAC PA 150875; 210: GMD NAC PA 150896; 211: GMD NAC PA 120591; 213: WD; 214: GMD NAC PA 190411; 216: GMD NAC PA 189764; 217: GMD NAC PA 190412; 218: GMD NAC PA 187562; 219: GMD NAC PA 190418; 220: GMD NAC PA 145205; 221: GMD NAC PA 150887; 222: GMD NAC PA 150877; 223: GMD NAC PA 150876; 224: GMD NAC PA 190419; 225: GMD NAC PA 150889; 226: GMD NAC PA 150897; 229: WD; 230: GMD NAC PA 190420; 232: WD; 233: WD; 236: GMD NAC PA 190410; 237: GMD NAC PA 150898; 238: GMD NAC PA 190402; 241: GMD NAC PA 189761; 242: GMD NAC PA 190416; 245: GMD NAC PA 190415; 251: GMD NAC PA 190401; 252: WD; 253: GMD NAC PA 120593; 254: GMD NAC PA 190422; 255: GMD NAC PA 190413; 256: GMD NAC PA 150893; 257: GMD NAC PA 120594; 258: GMD NAC PA 190414; 259: WD; 264: GMD NAC PA 150895; 265: GMD NAC PA 190409; 266: GMD NAC PA 150894 ; 267: GMD NAC PA 150883 ; 268: GMD NAC PA 150884 ; 271: KH; 319: WD; 320: GMD NAC PA 190399; 337: WD; 361: GMD NAC; 372: GMD NAC PA 145193; 373: WD; 382: GMD NAC PA 145218; 384: Canadian Natl Air Photo Library (NAPL); 390: WD

George M. Douglas

From the painting by E. Wyly Grier, R.C.A.

# Lands Forlorn

A Story of an Expedition to

## Hearne's Coppermine River

By

## George M. Douglas

With an Introduction by

## James Douglas, LL.D.

*With 180 Photographs by the Author, and Maps*

G.P. Putnam's Sons
New York                    London
The Knickerbocker Press
1914

The Knickerbocker Press, New York

# INTRODUCTION

M Y cousin, George Mellis Douglas, the leader of the expedition described in this narrative, is an engineer, and has had wide experience of small craft. After years of work in the arid Southwest he was naturally seized with an uncontrollable thirst for water, and one day told me of his longing to explore some of the rivers flowing into the Arctic Sea. Half in jest I undertook to "grubstake" him, if he would report on the copper–bearing rocks of Hearne's Coppermine River. He accepted the challenge. And this book is one of the results. It contains a narrative of travel in arctic Canada by three youths without native guides, who brought back scientific information of great interest and importance. Hearne's Coppermine River, flowing into the Arctic Ocean in Latitude 68° 49' N., and Longitude 115° 32' W., was chosen as offering an exceptionally interesting field for exploration, partly because the results might have a final commercial value.

In his introductory chapter the author describes how rumours and proofs of great copper deposits among the mountains in the far north prompted the earlier explorations of that inhospitable land. The Indian tales of the remote copper mines were as far from accurate as is many a mining report in the prospectus of a modern promoter. Instead of a mountain of copper Hearne found one lump of metal, which he picked up "among a jumble of rocks." Copper may once have been more prodigally exposed, but for centuries the croppings had been searched for float metal by the Eskimos and by the Coppermine River Indians, who used it as an article of barter over the whole North–west. It was carried by them eastward to the Hudson Bay and westward to the Pacific. The Eskimos are said still to prefer it for certain uses to iron.

Dr. Sandberg examined and mapped only the section of the Copper Range immediately west of the Coppermine River, but the party followed the same traps to the Dismal Lake and found them strewn on the shores of Great Bear Lake one hundred and fifty miles west of the Coppermine River. Simpson, Hanbury, and others picked up metallic copper on the beach of the islands in Bathurst Inlet. Thus these rocks, with a general north–east and south–west strike, have been traced for some three hundred miles, and if the same rocks really reappear in Victoria Land, where Stefansson describes the Eskimos as gathering copper, their cross–section must be vastly greater than the corresponding series on Lake Superior. The total area, therefore, within which copper ore may possibly be found, covers nearly ten thousand square miles.

Whether profitable ore occurs anywhere in this district can be determined only by a thorough survey, followed, should favourable indications be found, by exploratory development on an extensive scale. Our three explorers confirmed Sir John Richardson's diagnosis of the identity of these copper–bearing rocks with the melaphyres and conglomerates of the Keeweenaw series of Lake Superior. This was conclusively proved by Graton's petrological examination of the hundred samples of rocks brought out by our explorers. On analysis a large proportion of these rock samples contained traces of metallic copper—two of them were ores of profitable grade. What they did for the first time was to trace and map the succession of beds which compose this interesting—possibly commercially important—group of rocks. In this work they even preceded the Canadian Geological Survey.

The region may become one of the great copper producers of the world. It is now still inaccessible; but the easternmost exposure of these rocks, so far as known, is not more than five hundred miles distant from navigable water in Hudson Bay, over a possible railroad route. And a railroad from the south is already projected to within eight hundred miles of their most westerly exposure. As far back as 1845 Alexander Simpson anticipated that "It is possible that ere this century has passed, an organised system of internal communication such as that which traverses Northern Asia may place this valuable natural deposit within the reach of commerce." He was too sanguine. The climate is severe. But so is that of Lake Superior; and for about twenty–five years after the native copper mines of Michigan were first opened they were shut off from the east by ice from steam communication, for almost half the year.

As to fuel, lignite is plentiful within the arctic circle, and our explorers brought samples from a large bed of it which crops out on the shores of the Great Bear Lake. If copper ores therefore exist in large quantities, and are of a profitable grade, the unfavourable conditions are not sufficiently pro-hibitive to prevent their being exploited. And the race has happily not yet become so effeminate that men cannot be found willing to exchange the luxuries of civilisation for the novelties and adventures of work in the wilder-ness, when the inducement is added of winning from Nature the treasures she has hidden away in such inhospitable regions.

JAMES DOUGLAS.

# CONTENTS

# ILLUSTRATIONS

# Illustrations

# Illustrations

# Illustrations

# Lands Forlorn

I

AN HISTORICAL SUMMARY

E ARLY in the eighteenth century it was reported by the Indians, who
came to trade at the recently established Hudson Bay Co.'s post on
the Churchill River, that great deposits of copper existed near the
banks of a large river to the north.

At that time nothing was known of the northern coast of the continent, and a northwest passage was considered quite a feasible undertaking. The first attempts at exploration were made by sea, and they were
prompted by the hope of discovering such a passage and of finding mineral wealth at the same time.

Such expeditions were those of Knight in 1719; of Middleton in 1741;
and of Moor and Smith in 1746.

Knight and his crew perished to the last man, passing completely out
of knowledge for nearly fifty years, when traces of the party were at last
found on Marble Island, and a pathetic account of their tragic end was
gleaned from the neighbouring Eskimos.

The voyages of Middleton and of Moor and Smith resulted only in a
slightly more definite knowledge of the west coast of Hudson Bay. Wager
Inlet was found to be a landlocked bay and not, as was hoped, a possible

passage to the west. They also discovered Repulse Bay and Frozen Strait.

The chief result of these voyages was to demonstrate the difficulties of a northwest passage. The next attempt to reach the reputed copper mines, by the Hudson Bay Co. in 1769, was made overland.

Samuel Hearne was chosen to be in command of the expedition, if the word "command" may be used to express his complete dependence on a party of irresponsible Chipewyans. Indeed the organisation of this expedition shows a curious ignorance on the part of the Hudson Bay Co. of the real character of the Indians and an extravagant estimation of their influence over them.

Hearne's guides led him and his two white companions a couple of hundred miles out of the barrens, robbed them of everything they had, and then left them to find their way back to the post as they could.

Undeterred by these hardships Hearne made a second start a few months later. He took no white men on this occasion but only two of the company's Indian servants. As before he was dependent on a party of Chipewyans.

The thoroughly unreliable character of his guides probably would have doomed this attempt to failure also even if Hearne had not been obliged to return a second time, having had the misfortune to break his quadrant, an essential part of his outfit.

He could replace this at the fort only with an old and cumbersome instrument that had lain there for thirty years. Small blame to him that his subsequent observations were inaccurate.

On December 7, 1770, he made a third start under the guidance of a Chipewyan chief named Mattonabee. He was alone on this occasion and completely dependent on the whims of a party of Indians. The friendship of their principal man saved him from actual molestation, but he had no more influence with his guides than one of their own women.

The movement of the party was governed by their success in hunting. That they reached the Coppermine River at all was due to the decision of

Hearne's Chipewyans, who had been joined by a number of Copper Indians, to have an "Eskimo Hunt," evidently a popular form of amusement with the Northern Indians at that time.

Alexander Simpson, writing in 1845, says of these pleasant pastimes: "An Eskimo Hunt has always been a favourite diversion among the Border Indians. A decrepit Indian who acted as cook at a station where I resided often told me of the pleasure he enjoyed when an active youth in going on these expeditions."

Hearne's party was augmented till it numbered several hundred. They journeyed north and west to a lake afterwards called, by Franklin, Run Lake. The Indians left their families here; the final war-party consisted of sixty men. Travelling light and making long marches they struck the Coppermine River about thirty miles from the sea and followed it down to the last rapid.

It was by now the middle of July. A party of Eskimos was camped here fishing for salmon, as these people do to this day. The Indians waited for a chance when the Eskimos were quiet in their tents and then, in spite of Hearne's protestations, fell on them and slaughtered them all to the number of more than twenty. In memory of this massacre Hearne named the rapids the Bloody Falls.

The Indians had now accomplished the purpose of their journey and their only concern was to get back to their families as soon as possible. On their return journey Hearne examined what the Indians told him was the copper mine. He describes it as being about thirty miles from the mouth of the river and "an entire jumble of rocks and gravel." Only a small lump of copper rewarded his search. He was greatly disappointed; in his complete ignorance of the subject he seems to have expected that the copper really would be lying around in lumps like a heap of rocks as the Indians had reported.

Four hours was all the time he spent prosecuting the main object of his journey; with his band of Indians impatient to get home it was probably the most he could do.

After another year of roaming with the Indians Hearne finally arrived safely at Fort Churchill. He had made a journey that still remains one of the most remarkable on record. He was the first man to reach the shores of the Arctic Ocean and he discovered Great Slave Lake.

Hearne's unfavourable account of the copper mines and their undoubted inaccessibility killed all interest in them for a long time to come. The next account we have of these copper deposits was given by Captain Franklin and Dr. J. Richardson on their return from Franklin's overland journey to the Arctic, in 1819–20–21–22.

The main object of this journey was geographical research, but the reputed copper deposits were not overlooked, and Franklin was instructed "on his arrival at or near the mouth of the Coppermine River to make every enquiry as to the situation of the spot whence native copper had been brought down by the Indians to the Hudson Bay Co.'s establishment and to visit and explore the place in question."

Geographical discovery was paramount to Franklin, and the zeal of the whole party was directed to that end. They devoted only one day to an examination of the Coppermine Mountains and even that short investigation was made chiefly because they happened to kill some musk oxen in the vicinity which necessitated a delay in any case to dry the meat.

With such scant time at their disposal their investigations were of course very superficial, but so keen an observer as Dr. Richardson made good use of the time, and his description of the mountains is excellent in its accuracy (see Appendix A).

This is the first and hitherto the latest description of those great copper deposits. The difficulties of development and transport were considered so insuperable that the whole question was dismissed as being quite impracticable for any mercantile speculation.

The Coppermine River was used as a route by several explorers during the next thirty years, but nothing was added to our knowledge of the copper deposits till 1838 when Thomas Simpson called attention to the great abundance and extent of the copper deposits on the islands in Bathurst Inlet.

After Dr. Rae's journey to Wollaston Land in 1851 the country was unvisited till 1902 when David Hanbury ascended the Coppermine River. With two white men and some Eskimos he had made the extraordinary journey from Chesterfield Inlet to the Arctic coast, along the coast to the Coppermine River, returning to civilisation by the Great Bear Lake and the Mackenzie River. For daring and for extent this journey is almost unparalleled in the whole history of Arctic exploration.

Hanbury found copper float on the banks of the Coppermine River, but like Thomas Simpson his attention was chiefly attracted by the abundance of copper he found on the islands in Bathurst Inlet. He leaves the question of their commercial value to be decided by expert opinion.

## II

BY SCOW AND STEAMER. DOWN THE ATHABASCA AND MACKENZIE RIVERS

THE object of the present unostentatious expedition was to make a preliminary investigation of the Coppermine Mountains; to determine whether there was any analogy between these deposits and those of the Lake Superior district, and to decide whether the prospect was sufficiently promising to warrant investigation on a further and more comprehensive scale.

We considered that the success of this expedition would be best ensured by keeping the party as small and as efficient as possible. Three of us were indispensable; provided we could have got the right kind of man, we would have preferred four in the party. But the qualifications required in a fourth member were somewhat exacting, and the time at our disposal for choosing was short. We left civilisation decided to carry the work through with three men only. As will be shown, a fourth member did in fact join us eventually and at a time when his assistance was most opportune and valuable.

Our party consisted of:

*August Sandberg, Ph.D.*, of Sweden; chemist, metallurgist, and geologist. His travels in the remoter parts of Mexico had made him well acquainted with pack animals and life on the trail. He lacked training as a canoeist, but two canoeists were enough in the party and three might have been too many. His wide knowledge, his energy, his modesty, and his conspicuous unselfishness contributed greatly to the success and pleasure of the whole expedition.

6

*Lionel Dale Douglas, Lieut. R.N.R.*, a Canadian, sailor by profession, and with all the resourcefulness and ability to make the best of things by which sea–going men are justly distinguished. He was well versed in the handling of small craft, a canoeist of exceptional skill, an indefatigable hunter and sure shot with a rifle under all conditions.

*George Mellis Douglas*, a Canadian by birth, parentage, and early training; an engineer by profession, and by chance more than by any special qualification the leader of the present expedition.

We had decided that the best route to the Coppermine Mountains would be via the Athabasca and Mackenzie rivers, the Bear River, and Great Bear Lake, and thence by the Dease River and the small lakes on the divide to the Coppermine River. This was the route followed by the earlier explorers on their journeys to the shores of the Arctic: by Simpson in 1838–39; by Sir John Richardson homeward bound from his exploration of the coast between the Mackenzie and Coppermine rivers in 1827, and again twenty years later by the same explorer when he was returning from his search for Franklin.

Fort Norman, at the junction of the Bear River with the Mackenzie River, was to be our real starting point on our own responsibility. We had made provision to have our freight taken and to travel as passengers ourselves by the Hudson Bay Co.'s transports to that point.

Our plans for ascending the Dease River and crossing to the Coppermine River by way of the Dismal Lakes and Kendall River by canoe had been carefully thought out before we started, but they were quite indefinite regarding the crossing of Great Bear Lake and as to where we should pass the winter. We anticipated that the best plan would be to get a boat large enough to carry our entire equipment; to track it up the Bear River to Great Bear Lake, and to establish our own winter quarters on the Dease River. But this depended on several uncertain contingencies—on finding a suitable boat on our journey down the river; on being able to have it towed to Fort Norman by the Hudson Bay Co.'s steamer, and, finally, on finding an Indian crew at Fort Norman to help us track it up the Bear River.

Fortunately we were able to effect these combinations; had it been otherwise we would have made the journey by canoe all the way from Fort Norman, returned to that place in the autumn to spend the winter, and made an early start by dog sledge the following spring.

In one respect our little expedition differed from any other that has been made overland to the Central Arctic coast of North America. We had decided to be absolutely independent of native help, excepting only such aid as would be necessary to track our boats up the Bear River to Great Bear Lake. We had determined to be our own guides, hunters, and packers, and successfully carried out our purpose.

The Indians of the Mackenzie River valley have earned a most unenviable character for thorough unreliability and inefficiency. All travellers who have accomplished anything agree in describing them as worthless, shiftless, careless, unreliable, and generally contemptible.

Protracted residence in the country and a fuller experience of the Indians may disclose good points in their character, and mitigating reasons for the bad ones, but protracted residence in that country also lowers standards of reliability and efficiency and warps accuracy of judgment. Therefore it happens that the opinion of the man who has been longest in the country and who should know the Indians best must be received with especial reserve and caution.

It was advisable to go fully supplied with food for the total time we expected to be in the country, although we knew that the vicinity of the Dease River is fairly well supplied with caribou, and ptarmigan, and that fish abound in Great Bear Lake. The Indians have proved that one can live off the country, but it keeps them busy hunting and fishing from one year's end to another, ever following the movements of the game. They do not always succeed in living very sumptuously, and sometimes even not at all. For though the population of Great Bear Lake country does not exceed more than one person per five hundred square miles at a most sanguine estimate, even this scanty population is occasionally reduced to starvation.

We might have kept ourselves alive, but this would have taken all our time; and the movements of the game would not take us into that part of the country we wanted most to see. We expected anyway to eke out our supplies and knew that at times we would have to depend to a certain extent on successful hunting.

The calculations of supplies and their weights, the necessity for careful choice of equipment, and proving every detail of it, the worries caused by various delays, and most of all by one's own carelessness and forgetfulness— these things have been described so often by explorers that they may well be omitted here. Our equipment was perfect of its kind and proved so satisfactory in every respect that the actual experience of the trip showed scarcely anything that could be altered for the better, though we erred in taking an unnecessarily large quantity of certain supplies. We took nothing in the way of toboggans, snowshoes, or fur clothing, as we had been told by many acquainted with the Hudson Bay Co.'s posts on the Slave and Mackenzie rivers that those things could be got better in the country. This is quite wrong, such things can be had much better in civilisation. It was one of many instances to show how unreliable the advice may be of men who have been a long time in that country, and their commonly curious failure to appreciate the importance of time. We might, no doubt, have got together a good equipment for winter work if we had lived for a year at one of the posts before starting out and hired Indians to make snowshoes, toboggans, and to bring in dressed moose and caribou skins for moccasins and fur clothing, etc. By ransacking every post on our journey from the Athabasca to Fort Norman, we did in fact get together a sort of equipment for winter, picking up an old toboggan here, some snowshoes there, a capote and some moccasins at this post, some dressed caribou and moose hides at that. For although these things are the commonest necessities of life in that country, they are seldom obtainable when wanted offhand; the Indians make them only when they are actually required. After fuller experience and a more thorough realisation of the desperately hand–to–mouth way everything is done and provided for in the north, we had reason to consider ourselves lucky in getting together even this poor collection of essentials for our winter life.

Of our journey down the Athabasca and Mackenzie rivers I can give

The First Voyage of the *Aldebaran*

only a bare outline. To do it justice the subject would require a book to itself. It has already been described, and excellently well.[1]

We arrived at Edmonton on May 11, 1911. According to advices received from the Hudson Bay Co. the northern brigade of scows, by which

[1] Among many other good accounts of a journey down these rivers Agnes Dean Cameron's *The New North* gives an excellent idea of the country.

we were to travel, was due to leave Athabasca Landing the first week in June. Part of our equipment, including our two canoes *Polaris* and *Procyon*, and the more important part of our food supplies, had already been forwarded; our clothing, bedding, tents, arms, instruments, and ammunition we had with us; and the bulk of our food sup-

The Athabasca Stage

plies, hardware, etc., we intended to get at Edmonton. We were doing this in a leisurely way when, on May 17th, we were surprised by a telegram from the Hudson Bay Co.'s manager, who had just made a trip to Athabasca Landing, informing us that the brigade had *already left,* but that we might still overtake them by canoe at Grand Rapids, 150 miles down the river. Our canoes had gone with the scows, but the manager had kindly arranged that a large canoe should be ready for us at the Landing, expecting us to come by the next stage, arriving there on May 24th.

This spurred us to quick action. We had already added another

On the Road to Athabasca Landing

large canoe to our fleet, and on May 19th, we loaded all our stuff on two wagons, putting our latest addition, the *Aldebaran*, on top of one of the loads, and saw them start, fervently hoping that we might not see them again till we got to Athabasca Landing.

On May 23rd, we ourselves left by stage; this was not a single vehicle, but a regular convoy of wagons, and we numbered about twenty–five passengers all told.

Our own little party had been joined by Robert Service, who was making a journey to the North with the Hudson Bay Co.'s transport, and who,

"Eggie's"

like ourselves, had been surprised by the unexpectedly early departure of the brigade.

From Edmonton to Athabasca Landing is a distance of about one hundred miles, and the stage makes the journey in two days. On this occasion the roads were in fairly good shape, but plenty of evidence existed to show what they might be in bad weather. The first thirty miles of the journey is not an interesting country to travel through; it is a level, rich soil, with poplars and willows as the only trees. The country then becomes undulating with forests of spruce and jack pine, and some very pretty lakes. The land is being rapidly cleared and settled. At some of the settlers' homes we stopped for meals, and spent the night comfortably enough at a prosperous-looking farmhouse, delightfully situated. The next day's

Goodbye to Civilisation

journey took us over hilly country with worse roads. We passed a number of freighters stuck in mud–holes and one unfortunately rash automobile buried in black slime till it looked like some kind of a boat. We never knew in what plight we might overtake our own outfit, so were more ready in our sympathy than we might otherwise have been, but we finally rattled down the long hill into the Landing to find everything had arrived and all in good shape.

Athabasca Landing is a pretty little town situated at the top of a big horseshoe bend made by the Athabasca River, an emblem of luck particularly appropriate to this place, fortunate beyond most in its situation and prospects.

We were met by the Hudson Bay Co.'s agent at the Landing. He

On the River below the Athabasca Landing

entertained us hospitably and relieved our minds regarding our freight by assuring us that he could have it forwarded by a brigade of scows taking in supplies for the Roman Catholic Missions and about to start under the charge of Captain Schott, a famous riverman. These were due to arrive at Fort Smith before the Hudson Bay Co.'s steamer left that point.

A large canoe was ready for us and an Indian to act as guide, at least such was the thoughtful intention of the Hudson Bay Co., until we overtook the brigade.

Next day we spent a frantically busy time. Our outfit had to be assembled and the most indispensable things selected to make as big a load as possible for two canoes. It meant a lot of worrying work, unpacking, repacking, and general readjustment of our stuff, but at last the jobs were

*"Bannock Island"*

satisfactorily done and the two canoes loaded with about one thousand pounds weight in each. The balance we left ready to follow us with Schott's Brigade, trusting it would overtake us at Fort Smith.

The last letters were hurriedly written, and at 6.30 P.M., May 25th, we started on our long journey.

It had been a stormy day with high wind and frequent showers; a week of work and worry had culminated in our anxious efforts today. Now it came out a lovely placid evening, it was a blessed relief to shove out on that smooth swiftly flowing river and to leave all worries behind. It was no use to worry about anything now, all we had to do was to meet circumstances cheerfully and make the best of them.

Already we were far enough north to have long days and light nights.

Calling River

We kept on till nine that night, when we landed and made the first camp of the trip at an ideal spot in a thick grove of spruce. Nor did any of us suffer from the sleeplessness that the first night under the open sky sometimes brings.

Then followed two days of perfect weather and perfect conditions for canoeing; a steady, swift current that made paddling unnecessary except for the pleasure of it, an ever–varying scenery of lofty forest–covered banks, the dark spruce alternating with the bright green of aspen, cotton-wood, and birch in the first flush of early summer.

Our guide was a half–breed Cree; he didn't see any fun in paddling. He was worse than useless in camp and no good as a guide even if a guide's services had been necessary.

On the evening of the second day we arrived at the Pelican Portage;

Pelican Portage

a trader's store and a few log shacks built by Indians stand on a high bank overlooking the river. A couple of miles below this are the Pelican Rapids; our guide suddenly got nervous and said it would be necessary to get some one to pilot us down these.

While he was looking for some one at the houses, we investigated an old oil well, near the river, that had been drilled by the Canadian Government fifteen years ago. At 800 feet they struck gas, which blew their rigging out; this gas has been escaping and burning ever since. The casing had been reduced to a 2 $1/2$" nipple, the gas was under such pressure that it shot up six or eight feet before igniting, and threw up a flame twenty–five or thirty feet high, burning with a roar that could be heard a couple of miles away.

Our "guide" couldn't find a man, so we decided to be our own pilots. Lion

Grand Rapids

and I led the way with the *Aldebaran,* but the rapids proved only very swift water and easy enough at this high stage of the river; we made camp below them that night. From this point to Grand Rapids the river is very swift, the banks become higher and more abrupt, rising several hundred feet, usually in three or four well–defined wooded terraces divided by bluffs of sandstone and shale.

We reached Grand Rapids early in the afternoon of the third day; our average speed from Athabasca Landing had been about nine miles per hour, which gives some idea of the swiftness of the stream.

The Hudson Bay Co.'s brigade were still there camped on the island; the officers welcomed us as though we were long expected friends rather than more possibilities of trouble in the shape of passengers.

At Grand Rapids the river is divided by an island into two channels

The Lower End of the Island at Grand Rapids

nearly half a mile long, and in that distance the river makes a drop of about forty feet. The west branch is the main channel and the water rushes down that side in a terrific swirl. The scows land their loads at the head of the island and are run or lowered empty down the easier eastern channel and then hauled in to the lower end of the island to receive their cargoes again, which are carried over the island on a primitive wooden tramway.

The scows are built of 1" spruce, they are about 55 ft. long by 12 ft. beam and 3 ft. deep. They carry a load of about seven tons and are manned by a crew of five or six men. A large steering sweep, nearly as long as the scow itself, is their most characteristic feature. I measured one sweep forty–seven feet long and saw others that may have been even longer.

The brigade consisted of some twenty two or twenty four scows and

Scow Leaving the Island

a cook scow. There were about a dozen passengers including ourselves. Among them were two men sent by the Canadian Government to establish an experimental farm at Fort Simpson, who were taking their wives and families with them.

Only a few more scows remained to be run down the rapids; and a couple of days after our arrival we began the next stage of our journey.

From Grand Rapids to Fort McMurray is a distance of ninety miles, the river is very swift with nine or ten larger rapids. The scows can run all

The Athabasca Brigade

these but two, at which they have to be lightened and most of their cargo portaged.

It was a mode of travelling delightfully novel and interesting; a leisurely progress with frequent stops for meals or for adverse winds. A

The Cascade Rapids

very slight head wind was sufficient excuse to stop, in fact anything or nothing at all would bring the whole fleet to tie up along the bank for a "spell."

Nominally the brigade was under the charge of a captain, but its movements were in fact regulated by the whims of the Indian pilots, and we went ahead or stopped—mostly stopped—according to their inclination. Our voyage from Grand Rapids to McMurray took a week; this means that we were actually under way about two hours per day; it was a series of resting spells with short interludes of progress. But no one worried, tomorrow was as good as today, the weather was fine and bright, the scenery beautiful, and grub plentiful. The Indian crew had four, sometimes five, meals a day; the Hudson Bay officers and the passengers had three.

Cook Scow Running the Cascade

These meals were all at different hours, so the crew of the cook scow worked hardest of the whole outfit.

Moose and bears were seen occasionally as we drifted down the river; there were usually at least two or three rifles on each scow and the ensuing fusillade would do credit to a small battle. The apparent object of every Indian who had a rifle was to empty the magazine as quickly as possible, the range or even the visibility of the game had nothing to do with his shooting; if he only heard it crashing through the woods he would fire at the noise. The Indian may be a good hunter when he is alone, but when a bunch of them are together any game is fairly secure against damage.

We usually stopped for the night about 5.30 P.M. When there was

At the Cascade Rapids

any choice the Indians showed a lack of discrimination in choosing good camping places that was quite in character with their haphazard methods. From long custom one place was as good as another to them; when they felt like camping they camped, a better place three hundred yards farther on was nothing to their inclination.

Our scow journey was altogether thoroughly enjoyable; it was almost a matter of regret when we finally reached Fort McMurray. This was our first experience of a fur trading post, though it was then making its last appearance in that character and probably will soon be completely swamped by civilisation.

It is the centre of an area thought by many to be rich in oil, though the vast extent and depth of the famous Athabasca "tar sands" may be merely the remains of a once great oil field now uplifted and drained.

On Board the Scow

The Hudson Bay Co.'s steamer *Grahame*, by which we were to make the next stage of our journey, was awaiting us and we took up our quarters on board at once. The *Grahame* is the usual type of the shallow draught stern wheeler common on western rivers. She was a comfortable home, and nothing could be more pleasant than the truly hospitable and courteous way in which we were welcomed on board.

Fort McMurray is beautifully situated on a level point of land between the Athabasca and Clearwater rivers. It consists merely of a few log houses, a small store run by the Hudson Bay Co., and another by a Miss Gordon.

A crowd of Indians were camped on the point awaiting the arrival of the transports and the treaty–paying party. Here we made our first acquaintance with the dogs of the North; big, suspicious, hungry–looking brutes, with characteristically large strong feet. It was the idle time of the year

The Descent of the Athabasca River

for them, but they were none the happier on that account. They suffered from the heat, from flies, and above all from hunger; their food had been cut down to just about enough to keep them alive. Every teepee had three or four dogs stretched out in the coolest place or prowling around looking for something to chew, or else the chance of a fight. But everything chewable was hung out of their reach on tripods made of saplings, fights were much easier than food for them to find, and apparently what they enjoyed next best.

In some respects the next stage of our journey may have been more comfortable than the voyage on the scows; certainly it was far less novel and picturesque. To exchange a bed of boughs in the open for a small and usually stuffy cabin was of doubtful benefit, but this was offset by the com-

Fort McMurray

fortable saloon for meals, instead of the cramped table on the cook scow, where you sat on a sack of pork with the grease oozing through it, and thought yourself lucky if you could find room for your feet in a half empty sugar barrel under the table. I do not mention these small incidents of scow travel to illustrate its picturesqueness, but their novelty at least is undeniable!

We left Fort McMurray again a few days after our arrival. Below that post the river is stilt fairly swift, flowing through an alluvial plain with heavy forests of spruce, birch, poplar, and cottonwood. As the river approaches Lake Athabasca the shores become lower and the woods change in character until they are mere forests of willow covering low swampy ground. The nights were scarcely light enough as yet to permit travelling

A Trading Store at Fort McMurray

all the time; we usually tied up about 10 P.M. and started again at two in the morning.

The chief event of the voyage was our arrival and few days' stay at Fort Chipewyan. This was really our first acquaintance with a Hudson Bay Co.'s post, Fort McMurray scarcely counted. Fort Chipewyan is among the oldest and still one of the most important of all the fur trading posts in the North. It is also one of the most striking in appearance and situation; the Hudson Bay Co.'s buildings are dignified and harmonious;

they occupy a rocky point at the eastern extremity of the settlement. At the western end is the large Roman Catholic Mission. Between them a row of houses faces the lake. All the buildings are made of squared timber and all are whitewashed. With its background of forest-

The Hudson Bay Co.'s steamer *Grahame*

covered rounded and rocky hills the post makes a picturesque and pleasing appearance.

It is indeed a lovely spot in summer with the spruce–covered rocky islands in front and the great expanse of lake to the east, wonderfully beautiful in its constantly changing aspect. For at this time of the year the air is considerably warmer than the water; there is great refraction and all kinds of wonderful mirages; every minute makes some new and curious change in the appearance of the lake, and the distant

The Hudson Bay Co.'s Store at Fort McMurray

Idle Days

shore lines take all manner of appearances, or sometimes disappear entirely.

Fort Chipewyan was founded by the Northwest Trading Co. in the latter part of the eighteenth century; it is of historical interest as the base

The Clearwater River

and starting point of most of the famous explorers in the early part of the nineteenth century. It was an important post before Fort Dearborn was founded, and probably appeared very much the same to Sir John Franklin in 1820 as it does to us; equally little change has taken place in the general life and character of the Indians; the descriptions given by Franklin, Richardson, and Back apply today. Time has dealt gently indeed with it, the islands are as inviting, the lake lovely as ever in its constant changes, and the skies show the same ineffable serenity, while Fort Dear-

The *Grahame* Tied up for the Night

born has changed in name and nature. Who shall say the change is for the better?

We left Fort Chipewyan on June 13th. The scenery on the Rocher and Slave rivers from Athabasca Lake to Smith's Landing is in some re-

The Hudson Bay Co.'s Buildings at Fort Chipewyan

spects the most beautiful of all that long voyage; rocky shores alternate with alluvial banks, and there are many picturesque islands.

Our arrival at Smith's Landing, the northern limit of the *Grahame's* voyage, seemed a matter of most complete unconcern to the inhabitants of that pretty little village. We had been there some time before any one showed enough interest to come to the steamer, even the very dogs treated us with a chilling indifference.

From this point to Fort Smith, a distance of sixteen miles, the river is a succession of rapids; we estimated a drop of about 125 ft. in that distance. Some of these rapids are of a very formidable description; at the Cassette Rapids the river is about three-quarters of a mile wide and the rapids are a most impressive sight.

The Roman Catholic Mission at Fort Chipewyan

We spent several days at Smith's Landing, and saw the balance of our supplies, which we had left at Athabasca Landing, arrive safely by the Roman Catholic Mission steamer.

Most of the freight is taken over by wagon, some of it is run down in scows, which make portages at the worst places. We saw all our stuff, including our canoes, despatched by wagon, then walked across ourselves; the road is a pretty one through dense woods of small timber and for the most part over level sandy ground, though there are some bad marshy

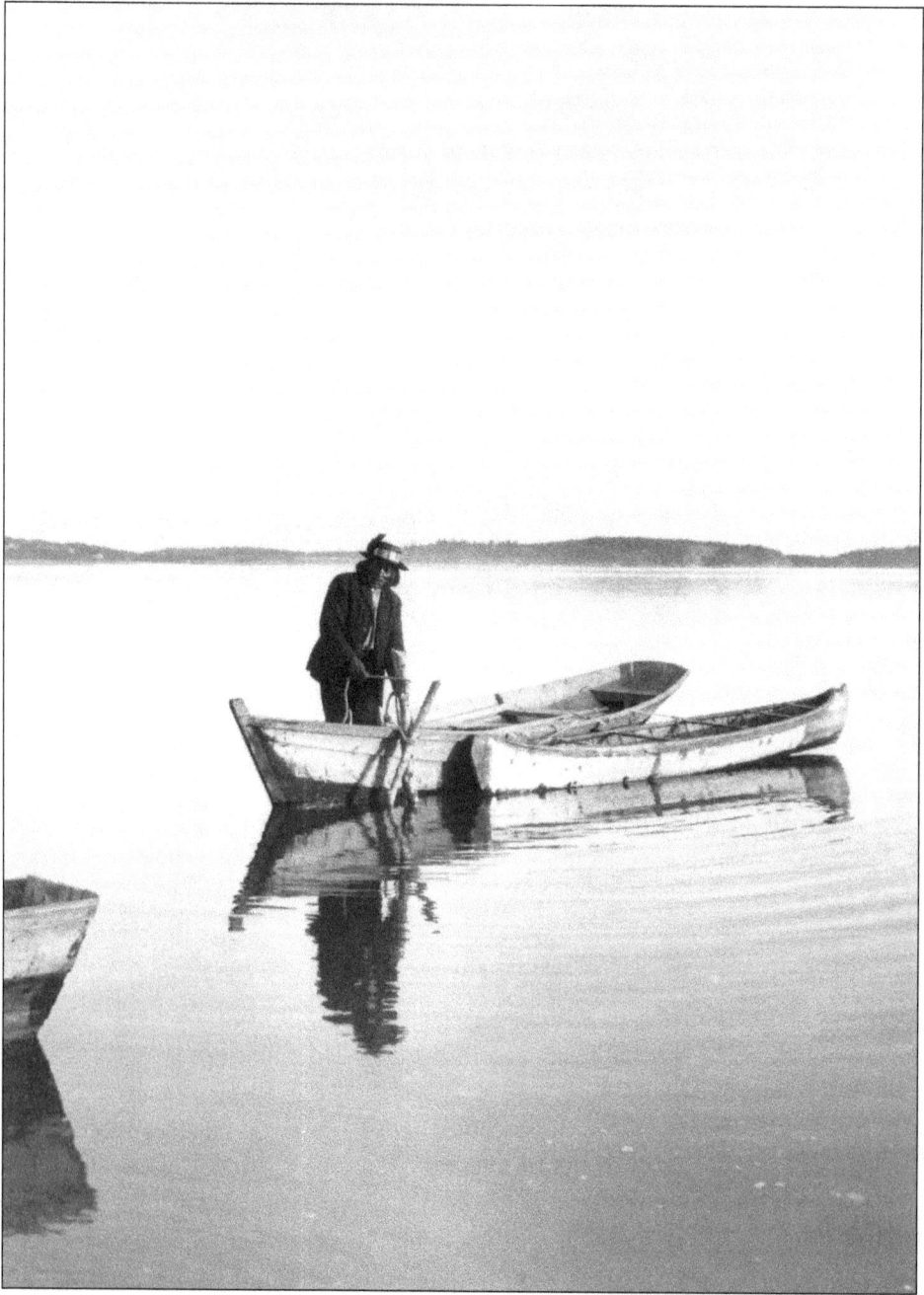

Coming from the Nets

stretches. For mosquitoes and flies Fort Smith is probably the worst place in all the world; a fine view of the river is about its only other claim to distinction. The road is practically level from Smith's Landing to Fort Smith, but while the Landing is at the water's edge, the Fort is about 130

The Eastern Part of Fort Chipewyan

ft. above it, built on a level sandy plain which ends abruptly in a steep bank. The constant erosion of the river continually undermines this sandy bank and causes tremendous land slides; in some places acres of the plain have slid away and by its subsidence whole forests of trees will be reduced to strange and unnatural angles.

The Hudson Bay Co.'s steamer *Mackenzie River* was tied up at the river bank and we were welcomed on board by the genial Captain C. S. Mills. This steamer was our home for the next three weeks, a comfortable well–appointed boat and her captain a truly courteous and obliging host, whose versatility and great experience of the North made him an uncommonly interesting companion. The *Mackenzie River* had been built to Captain Mills' design and under his superintendence, a difficult task in a

The Outskirts of Fort Chipewyan

place so inaccessible to labour and supplies as Fort Smith. Especially does her builder deserve great credit for the design; which has to meet such different conditions as the vast inland sea of Great Slave Lake, requiring stability and free board; and the swift, in some places shallow, Mackenzie River, making light draught a necessity.

A fortnight after our arrival at Fort Smith we were ready to proceed, the cargo had all been transferred, our own stuff was at last all together, and on June 28th we made a start.

The nights were now so light that we could run without reference to the time. Only stars of the first magnitude had been visible lately and

*"Bull Dogs"* on Cabin Window

then only at midnight. Even to these we now bade farewell; it was next autumn before we saw them again.

The lower Slave is not an interesting river, it flows through a densely forested alluvial plain, the banks are sandy and gradually got lower as the Great Slave Lake is approached.

We reached Fort Providence the second day; this is a small village of log houses built on the flat shore of a bay in the lake. It is not a pretty place even in summer, and in winter it must be unspeakably bleak.

Fort Resolution is another post well–known as the base of exploring expeditions. It was of particular interest to ourselves as the place where we succeeded in getting most of our equipment for the winter, including

Fort Smith

some rather fine snowshoes. We had a delightful trip across Great Slave Lake, with fine weather all the time. Especially was the clear blue water a most pleasant change after the muddy rivers we had been descending.

The post of Hay River, situated near the mouth of the river of that name, was our next port of call; we arrived there at 2 A.M. and spent a couple of hours only. It is interesting chiefly on account of the Protestant Mission, a mission whose methods and organisation deserve the highest praise.

At noon on July 1st, we entered the Mackenzie River, a noble river indeed. Most of the great rivers of the world flow through low lying country, for instance the Amazon, the River Plate, and the Mississippi. The St. Lawrence is an exception, but the Mackenzie is a far more notable one; the scenery below Fort Simpson is in-

The Hudson Bay Co.'s Steamer *Mackenzie River* at Fort Smith

comparably grand, a mighty river flowing among mighty mountains.

Fort Simpson was our principal port of call on the Mackenzie River, though it is of far less importance now than some years ago when it was the chief post and headquarters of the Hudson Bay Co. for the Mackenzie River district. There is a general air of listlessness, decay, and departed greatness about the place.

The post is situated on a very fertile island at the junction of the Liard with the Mackenzie River. The farming outfit was landed at this point to begin their experimental work. Such results as they may obtain will be of value only as far as that small alluvial island is concerned. It would be entirely misleading to apply them to the surrounding country, which is quite different in character, and even a few miles away from the river, in climate also.

On the Lower Slave River

Part of the farm outfit had been loaded on a York boat and was towed from Fort Smith by the steamer. We decided to buy this York boat and Captain Mills consented to tow it the rest of the way to Fort Norman for us. It was thus that the *"Jupiter"* as we called her came into our possession; we had looked on her as a possibility ever since we first saw her hauled up on the beach at Fort Smith. She was simply a big open boat over 50 ft. long and 12 ft. beam amidships, 3 1/2 ft. deep, very straight in her lines. Such primitive rigging as we wanted we were able to get here; we made a sail out of an old scow tarpaulin 18 ft. by 22 ft. We also bought the floor of an old house that had been pulled down and two small windows; these were very valuable to us later on. The picture shows the *Jupiter* loaded up with the farm outfit. The canoe in the foreground is the Radford Expedition starting on their

Fort Resolution

trip which had such a fatal ending. The man in the bow is Radford and Street is in the stern.

Our pleasant voyage came to an end only too quickly. Fort Wrigley was the last port before arriving at our own destination; this is merely a few small log shacks and a diminutive chapel.

On our arrival at this place we learned some rather gruesome news. Two white men had brought in a small outfit and had built a shack on the Salt River, a point about a hundred miles below Fort Wrigley where they had spent the winter trapping. An Indian passing the place about a month previous had found them both dead in their shack. Everything had been left strictly undisturbed until the arrival of the Northwest Mounted Police Inspector on the steamer, so that a proper investigation could be made.

We left Fort Wrigley in the forenoon and arrived at the Salt River that same evening. The trappers had chosen a very pretty spot for their

Fort Simpson

shack, with a fine view up the Salt River and across the Mackenzie. It was the usual small log cabin about 12 ft. by 14 ft. The men lay dead in their bunks, one with his head a shapeless mass, blown out of all resemblance to anything human by a soft point bullet from a high power rifle. On a small table beside the bunks lay a dirty note–book and a bottle containing a little carbolic acid.

The stench was insufferable, worse than any other form of decomposing animal matter, and blended with it was the peculiarly acrid smell of old smoke from spruce fires. One could remain in that loathsome atmosphere only a few minutes at a time; the bodies were in a state of decomposition so advanced that it was necessary to break the bunks down and carry them out as they lay. Close to the house on that pleasant point we buried them both in one grave, dug as deep as the frozen ground permitted.

The Radford Expedition. The *Jupiter* in the Background

In the note-book we were able to make out the following message written on different pages and evidently at different times.

"Cruel treatment drove me to kill Peat. Everything is wrong he never paid one sent ship everything out pay George Walker $10. . . . . I have been sick a long time I am not Crasey, but sutnly goded to death he thot i had more money than i had and has been trying to find it.

"I tried to get him to go after medison but Cod not he wanted me to die first so good by."

"I have just killed the man that was killing me so good by and may god bless you all I am ofle weak bin down since the last of March so thare hant no but Death for me."

He had shot the other man and then probably ended his own life by a dose of carbolic acid.

Trappers' Shack on the Salt River

Behind the shack, farther up on the hillside, was a small log storehouse; there were a few sacks of flour inside and a fine collection of furs that had been accumulated by these men on their winter's trapping. We took them, and the rifles, and such evidence as there was and held an enquiry at once as we proceeded down stream.

It was late by the time this was finished, and after that the purser and I had another and more complicated job getting our accounts straightened out.

Twilight was giving way to bright day by the time we had finished, the lofty Bear Rock below Fort Norman was already visible, clear and pure in the morning sun. But the night had been a harassing one, and we would need all the energy possible this day; close as we were to our journey's end we turned in to take what sleep we could.

III

THE VOYAGE OF THE "JUPITER"—THE BEAR RIVER AND GREAT BEAR
LAKE

SELDOM in my life have I heard a sound so unwelcome as the fore-boding scream of the whistle that jarred me out of the first beginning of a heavy sleep. We had arrived at Fort Norman; it was 2.30 A.M., that hour when one's courage is at the lowest ebb. The Hudson Bay Co. with their kindly help could do no more for us; the time had come for us to shoulder our own responsibilities, and to bid our kind hosts and pleasant travelling companions farewell.

The steamer stopped at Fort Norman for only a few hours, our stuff was dumped out on the beach, the canoes landed, the *Jupiter* cast loose, and the *Mackenzie River* stood out in the stream again and was soon lost to sight on her long voyage still farther north.

We were on an exposed shore, the first job was to get the *Jupiter* in a fit state to put our stuff aboard. Since leaving Fort Resolution she had been the home of a team of dogs and was in a condition indescribably filthy; it took several hours' hard work scrubbing, swabbing, and baling to get her reasonably clean and then we loaded all our stuff on board. It was after-noon before we finished; we had been toiling all day in a hot sun, not even taking time for lunch; tormented by flies, and almost overpowered by the smell of our new home. For though I said reasonably clean, that was merely to the eye; the smell didn't evaporate for hours. The last twenty-four hours had been a succession of bad smells. At last we got everything

48

loaded and tracked the *Jupiter* to a sheltered spot about half a mile up the Bear River; only then were we at liberty to make an acquaintance with Fort Norman.

The post is the usual small village of log shacks; a small store run by

The Beach at Fort Norman

the Hudson Bay Co. and another by the Northern Trading Co. There is a Roman Catholic Church and Mission, also a small Protestant Church. The latter was out of commission and shut up when we were there. The post is beautifully situated on a point between the Bear and Mackenzie rivers, and the outlook is very fine. In front is the great Mackenzie River, and far beyond it the distant Rocky Mountains; to the north is the lofty Bear Rock with its variegated colours.

We made the acquaintance of Mr. Leon Gaudet, the Hudson Bay

Co.'s factor, and of Hornby and Melville, two Englishmen who had been several years on Great Bear Lake hunting and attempting to trade with the few Indians who live in that desolate country. They had spent the last winter near old Fort Confidence in Dease Bay, where they had built a

Main Street, Fort Norman

small house. Melville was on his way out to civilisation again, Hornby had decided to spend another year in the country. The preceding summer they had met the Eskimos from Coronation Gulf, the last remaining primitive people on the continent quite untouched by civilisation and still living in their primitive way. These Eskimos, who have lately been made famous by a sensational title, come inland in the summer from Coronation Gulf and the lands farther north to hunt caribou and get wood for sleighs, etc. Hornby and Melville had met them on the edge of the Barrens to the northeast of Great Bear Lake. The Roman Catholic Mission had decided to send one of their fathers to get in touch with these Eskimos and attempt their conversion. Father Rouvier, O.M.I., had been well chosen for this hazardous undertaking; and Hornby, who wanted to do

Mackenzie River and the Bear Rock from Fort Norman

some fur trading with them, had arranged to accompany him on his journey.

We learned also that Mr. J. Hodgson, a retired Hudson Bay Co.'s factor and his family, had also spent the preceding winter on Dease River, trapping and hunting caribou. They were to leave Dease River as soon as the ice in the lake broke up, returning in a York boat left there by Hornby and Melville, and we thought it possible that we might meet them on their way home.

Our first concern was to get together an Indian crew to track our boat up the Bear River. Most of the Bear Lake Indians were at the post and would soon be going up the river again to their hunting grounds on the lake. By the aid of the Hudson Bay interpreter we negotiated with these

The Bear Rock from our Camp

people, but could not induce them to undertake the job. Besides a large number of birch canoes they had a small York boat that they had obtained from the Hudson Bay Co., and their own outfit probably required all hands. Even the offer of our York boat as soon as we got our stuff across the lake did not tempt them, "for that" they said "cannot be divided among us, only one man can own it." Another specific objection was the pants we wore; innocent Duxbak pants, but they saw an unfortunate likeness to the uniform canvas pants worn by the R. N. W. M. P., and they didn't want any of them in their country. For several days we treated, entreated, and negotiated with these people and under the disadvantage of having to carry it on through an interpreter, but there was nothing doing. Meantime we had a fairly comfortable camp at the place where we had taken

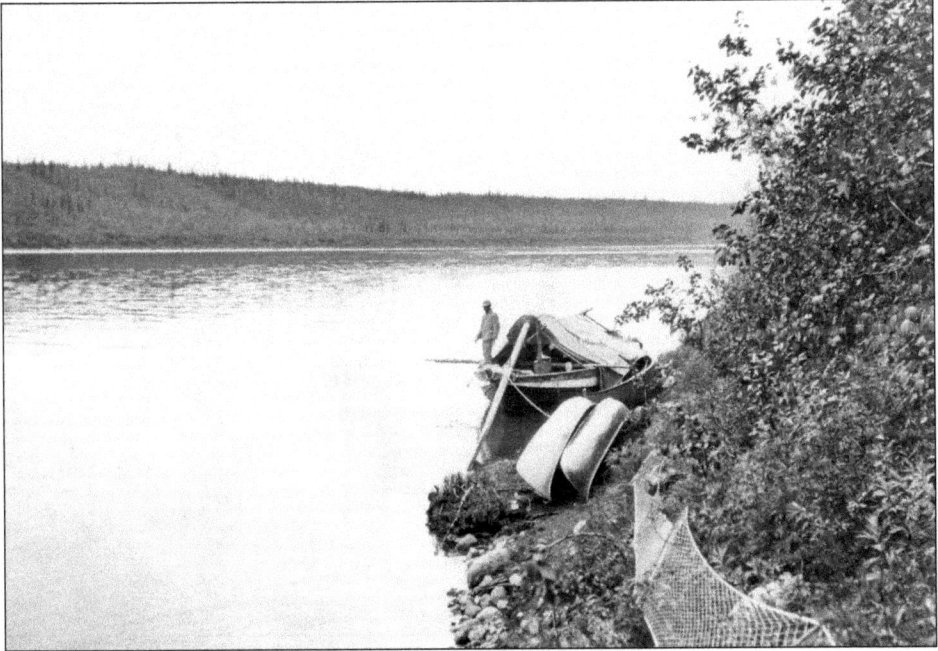

The *Jupiter* at Fort Norman

the *Jupiter*. The water of the Bear River is brilliantly clear and was then at a temperature not far from the freezing point. Hornby and Melville were camped on the point between us and the post, a place that became familiar enough to ourselves more than a year later.

It was an irksome and exceedingly trying time; we were impatient to start on our voyage, but it looked for a while as though the *Jupiter* would get no farther than her present anchorage. We even began preparations for making the journey by canoe, when by the exertion of considerable trouble and personal influence, and the offer of high wages, Mr. Gaudet succeeded in getting together a crew of local Indians for us.

Six men were all he could get; including ourselves this made a short-handed crew to track a York boat up that swift river, but our load was

Lion and the Doctor

fortunately not a heavy one for the size of the boat; we had 3 1/2 tons of stuff all told, and the *Jupiter* drew about eighteen inches of water. Four or five inches of this was keel, and a confounded nuisance that keel was to us before we got to Bear Lake, and very little use after, except perhaps that it afforded extra strength to our ship on one occasion when it was badly needed.

Saturday, July 8th, was the day fixed for our start. We decided to leave the *Procyon*, one of our canoes, at Fort Norman, as it might be necessary for us to leave the country by way of the Porcupine and Yukon Rivers. We saw her safely bestowed in the care of the Hudson Bay Co., said good-bye to our friends at the post, and returned to our ship to await the crew. By 3 P.M. they showed up in a big birch canoe that they brought to return

Our Indian Crew

to Fort Norman in when we got to Bear Lake. Their names, or at least the names by which we knew them, were Lixie Trindle, Clement, Samuel, David Wright, Horatio, and François. Lixie Trindle was the only one who knew a little English. He was the captain of the outfit till we got to Bear Lake, and although he had not been given a particularly good reputation we found him a reliable and a hard worker. Samuel was the humourist of the party; he had a broad heavy face with a singularly fatuous expression. We never knew what his jokes were about, probably they were at our own expense. François was a small well–built man, very strong, very quiet, and a hard worker so long as he had an example. He had his wife, a little girl, and one dog with him, and he wanted to cross Great Bear Lake, with us, and to work for us a couple of months, which was convenient

We Make a Start

enough. His wife could talk a little French and was afterwards the only medium by which we could communicate with the Indians. The rest of the Indians were quiet, hard–working men, and though we were short handed we certainly made up for it in the quality of our crew. The opinion we had formed of the Northern Indians, generally, was certainly improved by our small personal experience with these men. But they were no doubt better than the average Indians; moreover the character of the work, a short severe effort with the end in plain view a few days ahead suited them better than a long trial with its end indefinite, and requiring persistence and steadiness of purpose.

The first thing our crew did was to start a fire, put on the kettle, and settle down to a good square meal and a smoke. Then the grand start was

Banks of the Bear River

made at last; we shoved across the river to the north shore and were soon under way in harness.

Lixie took the big steering sweep; this was his job all the way up the river. At this part of it four men on the tracking line were enough, and we took "spells" on the boat. Lion and I were among the idlers at first and the Doctor was on the line.

It was a dull cloudy day but pleasant enough sitting in the boat watching the shore slip past and feeling that we had got fairly started on our voyage. The great rise of water and the breaking up of the ice in spring keep the banks clear of trees. For the first forty miles the shores of the river are generally pleasant grass–covered slopes, at that season profuse in all kinds of flowers; roses, violets, fireweed, wild onions, and all in bloom at

Bear River — The Franklin Mountains

the same time. Behind and above was the unbroken spruce forest, much denser and the trees much taller than I expected to find them here. Altogether it is a most beautiful river with its brilliantly clear water, its charming banks, and the fine views of mountains in the background; Bear Rock was then behind us to the west, and the Franklin Mountains to the east ahead of us.

The river is very swift and generally shallow; except near the mouth where the river bottom is limestone, and at the head of the rapids, we saw no rocks in place; it is all gravel, or gravel and boulders.

The gravel bars and shallow places soon began to give us trouble; the *Jupiter* was frequently ashore, and when this happened it was the duty of the crew "resting" aboard to jump into the water and heave her off, no pleasant job in that icy stream. Lion and I were pretty lucky in this

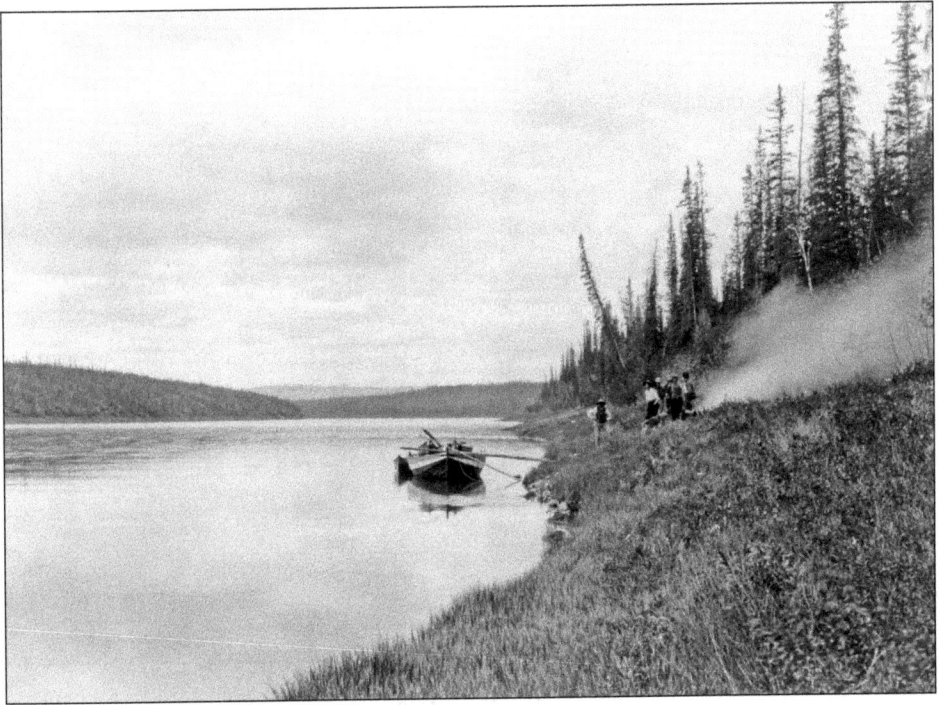

A Camp on the Bear River

respect, it didn't happen very often to us, but from early experience we were handy with a pole and one of us was always on the lookout in the bow.

The Doctor was the first victim, his spell of resting happened to come at a bad stretch of the river and he spent it for the most part up to his waist in water and his back under the boat struggling to shove her off gravel bars or boulders. I wondered at his apparent hardihood and indifference to the

cold, but soon found the explanation of it. A small stream called the Willow
River flows into the Bear River about ten miles above its mouth. We had
noticed this yellow coloured water along the north side, well defined from
the clear water of the Bear River, but I did not suppose there was more

Mount Charles

than 20° Fahr. difference temperature between them! But this condition
lasted only for the first evening.

At one stretch it was a hard job to get the boat up at all; I began to
wonder how long our crew would stand the strain, but I did an injustice to
their perseverance. To add to our troubles and discomfort it came on to rain
just before we made our first camp below the Willow River; a sloppy camp
it was, we were soaked through, from both above and below, but thanks
to the abundance of dead dry spruce one always finds in these north-

ern woods, we soon had a big fire blazing, and a good supper of bannock, bacon, and tea. We put up one of our canoe tents that night, it was a tight fit for three of us; the Indians made shift with their mosquito bars and a tarpaulin. It rained heavily during the night.

Ice Field below the Rapids

Next day things looked more cheerful, the weather cleared up, the conditions for tracking were better, and we made good progress on our way up the river. It usually took about an hour and a half to get breakfast and break camp in the morning. We were under way at 7.30 A.M. that day and kept at it till 11 P.M., stopping three times for meals and taking turns on the tracking line as before. The banks were of the same pleasant character; we walked among grass and flowers all day; the mountains ahead gradually got higher and more distinct,

Ice Cliffs at the Rapids

while behind us the Bear Rock ever diminished and became more and more ethereal.

The surroundings were lovely but the work was hard; a slavish task, hauling in harness like a very mule. During one of my "spells" aboard,

Along the Ice Cliffs

I noticed a dead caribou floating near the shore; it had escaped the notice of the Indians and my own attention had been called to it by the smell. Very thoughtlessly I pointed it out to Lixie; he stopped the boat instantly, and the Indians held a grave council over that stinking carcass, testing strips of meat and doubtful for a while whether they would take it along or not. Finally, to our unmitigated relief, they decided not to!

The mountains ahead of us were a spur of the Rockies called the Frank-

We camp at the Rapids

lin Mountains; the Bear River cuts right through them, and the rapids are at this point. We camped about ten miles below the mountains tonight feeling that we had done a good day's work.

Next day the navigation was more difficult; the river is full of islands

Taking a Spell

and shallow bars, the current was swifter, and all of us were needed on the line, so there were no resting spells for any one.

By four in the afternoon we came to a great field of ice on the south side of the river just below the mountains. We could walk over the top of this and track the boat along the edge, but at the rapids themselves the ice was piled up in great masses all along the shore, forming a rugged, unbroken wall for several miles. The prospect looked bad enough; it was impossible to track the boat from the top of those ice cliffs, and a decidedly risky proceeding, to walk along the bottom of them; but we had no choice in the matter, this was the only way we could get on, so we got out our heaviest tracking line and made the attempt.

Sometimes there was a little beach, enough to give us a good foothold,

The *Aldebaran* and the *Polaris* being Tracked up the Rapids

at other times we had to struggle along up to our thighs in that swift icy water with a whirl of rapids on one side and a sheer wall of ice on the other, often with precariously balanced overhanging masses of ice above our heads. We were too busy desperately hauling on the line to think what the consequences might be if any of these threatening pieces gave way while we were below.

We camped that night right alongside the ice in a small cove, a mere depression in the line of the cliffs, but it gave a certain amount of protection to the *Jupiter,* and the wall was sufficiently broken so that we could cut steps in the ice and reach the top of the cliffs.

The Indians passed the night ashore, the rest of us slept on the *Jupiter;* the ice gave us a good protection against the mosquitoes at any rate.

A Disheartening Prospect

We had turned in feeling fairly confident, but when we came to start again next morning, we found that our troubles were only beginning. We had made scarcely a hundred yards from our camping place, and that only by half an hour's hard work, when the water became so swift that we could make absolutely no further progress in spite of ill our efforts. We decided to lighten the *Jupiter* and track some of the stuff beyond the rapids by canoe, so loaded up the *Aldebaran* with about 1100 lbs. and the *Polaris* with 600. This was all the load they could safely hold in that rough water.

Five of the Indians took them up the river, one steering in each canoe, one tracking the *Polaris* and two the *Aldebaran*. I watched them start with much misgiving and anxiety; this day of waiting was the worst of all our voyage on the *Jupiter*. The prospect ahead was very discouraging, and

The Last Stretch of the Ice Cliffs

the idleness gave us plenty of time to realise the difficulties and dangers. A stretch of unbroken ice wall nearly a mile long still remained to be passed; beyond the ice, and as far as we could see, were high cut banks, equally forbidding and formidable. The bright breezy day and clear blue sky seemed only an additional aggravation; worst of all was the constant roar of the rapids. At four in the afternoon the Indians returned; they had taken the canoes safely about four miles up the river, but reported the tracking to be very bad. We had a good feed all around, the inevitable *première pas* of the North, and then got into harness once more for a supreme effort.

Lightened by nearly a ton of weight (for the canoes as well as the cargoes they had taken had been on board) the *Jupiter* came along easier and

Cut Banks above the Rapids

without extreme effort, and we got her up a part of the rapid that had been quite impossible before. But we had a hard struggle before we finally passed the ice. Then came cut banks with difficult footing; presently, and worst of all, several hundred yards of loose sliding rock. The continual erosion of the river made this particularly dangerous; the water is deep and swift right to the shore; we had to be most careful of our steps not to bring down an avalanche of rock that might sweep us all into the stream. The Indians were particularly nervous at this point, and once a small slide did take place that pretty nearly demoralised them. But we got up safety at last and made camp that night at the place to which the Indians had taken the canoes, devoutly thankful that all had gone so well and feeling that we had got over one of the worst parts of the journey.

On the Upper Bear River

The next day broke fair and tranquil, the whole aspect of the river had changed, and we had pleasant grassy banks again and good going. It was inexpressibly grateful after the strain of the last two days; we thought tracking almost a positive pleasure that fine bright morning as we swung along among the thick grass and flowers with a lively recollection of our late experiences among ice and rock slides. But this feeling of elation soon wore off under the incessant toil.

On the afternoon of the third day after leaving the rapids, we sighted

Great Bear Lake; never did open water look more welcome to us than that calm expanse, wonderfully blue, serenely peaceful. Head winds might come, storms or shoals, but here at least there would be no more of that incessant slavish hauling.

Our First View of Great Bear Lake

The upper stretches of the river had been so swift as to require the whole crew on the line all the time; the last two days in particular had been very fatiguing. The unusually hard work and hasty indifferent meals were telling on us all, and although they had worked beyond our expectations, we were getting decidedly tired of the Indian crew and looked forward to the better organised way of doing things that we could introduce as soon as we got on the lake.

At last we got to a point where the tracking line could be coiled away. We got up the mast and sail in a makeshift manner and stood across the end of the bay to Sir John Franklin's "Little Lake" where we knew that we would find a good harbour to refit the *Jupiter* for her voyage across the lake. It was on the shore of this lake that Sir John Franklin established his winter quarters on his expedition to explore the Arctic coast in 1825-26-

Entrance to Sir John Franklin's *Little Lake*

27; the place was named Fort Franklin. Fort Norman, that post so often moved, was also located here for several years. It was built on the low southern point at the entrance to Little Lake. The houses shown in the picture, now inhabited by the Indian, Johnny Sanderson, are probably remains of the post.

It was a fine, sunny afternoon, and we had a light fair wind, a pleasure indeed after our late toils to sit at our ease and feel the boat slipping along quietly through that clear water. It clouded up from the north, and we were suddenly struck by a head squall that obliged us to lower our sail and take refuge under a low rocky point, but the wind soon died away to a calm and we made Little Lake under oars. This is a pretty sheet of water about a mile long and more or less half a mile in width; it is really the wide expansion of a small river and is connected to Bear Lake by a short, deep,

At Little Lake

and narrow channel. A number of Indians, the families of those Bear Lakers we had seen at Fort Norman, were camped near here, and at a little settlement of log shacks about a mile farther along the lake shore; their nets were set off the mouth of this river and in the short channel to Little Lake; some of the women soon came along to our camp in their small birch–bark canoes. We tied up right alongside the shore and made camp just below the site of Franklin's old house, though of this there are no remains but a pile of rocks that had once been a chimney. It was a good harbour and a good place to refit, but in all other respects a poor camping ground. The only trees around were small larch and spruce; firewood was scarce and mosquitoes were plentiful.

The Indians decided to start back to Fort Norman that same night, so I wrote the last letters that would reach civilisation for more than a year, wondering at the time whether they would be received with due appreciation and judged with due leniency. I was completely tired out, the mosquitoes were in clouds, and the Indians impatient to start; altogether those letters were written under very trying circumstances.

With Lixie acting as interpreter we made a final arrangement with François, who agreed to work for us for a couple of months.

It was one in the morning when the Indians finally started back to Fort Norman in the big birch canoe we had brought along with us. I was just acting ready for a big sleep when I noticed the water was gaining so much in the *Jupiter* that some of our stuff was in danger of getting spoiled. I had to turn to and bale her out at once and pile some of the cargo ashore and the rest out of the way of the water; it was 3 A.M. and the sun well above the horizon before I got to bed.

We had always agreed that our first day at Great Bear Lake should be a day of rest; it proved instead a day of considerable activity. I woke up at six, anxious about the *Jupiter*, and found her leaking so badly that I called François out and we unloaded most of the remaining stuff before breakfast. While we were having this we saw the York boat of the Bear Lake Indians crossing the lake. This explained the anxiety of our Indians to return and some of their zeal in working so hard on the way up the river. Unknown to ourselves it had been a kind of a race; with their smaller boat and bigger crew the Bear Lakers had been boasting that they would beat us to the lake, and our crew evidently wanted to exult in their victory to the utmost by passing the others on their way back to Fort Norman.

We put in a busy day; we unloaded all the cargo and caulked the various leaks, then restowed everything in the best shape for our voyage. Lion changed the sail; we turned our former mast into a yard and sacrificed our big steering sweep, making a mast of it; we used a rudder when sailing.

Lion's experience with square rigged wind–jammers was valuable indeed on this occasion. By six that evening we were ready to start, but the light fair wind died away so we decided to wait.

Lion and I paddled up to the little Indian settlement and saw Father Rouvier and Hornby, who had come up the river with the Indians. They were all going on to Dease River, the Indians to hunt caribou, while the father and Hornby intended to get to the Dismal Lakes if possible and establish quarters for the winter there. We got back to our ship early and that night I had one of the best sleeps that had ever fallen to my lot.

Next day, Sunday, July 16th, was a lovely bright day with a fresh fair wind. We made an early start, poling the *Jupiter* out through the channel, then we hoisted our sail and stood across to the settlement, where François and his wife wanted to go ashore. We hove to waiting for them, impatient at the delay and the fine fair wind we were wasting; at last they came off in the *Aldebaran* with another little girl and another dog! While waiting for François, Hornby came out in a boat to see us; we said good-bye to him again, hoping to meet next time on the Dismal Lakes.

We had things properly organised on board now. The *Jupiter*, as I have said, was about 55 ft. long by 12 ft. beam, like a big open skiff. The mast was stepped amidships with fore and back stays and shrouds, the single sail we carried was 18 ft. high by 22 ft. wide, we steered with a long tiller from a small deck aft; we had a good spirit compass and this was set in chocks on the after–deck. The cargo was distributed so as to leave one clear space just forward of the mast, another just aft, and a third below the stern deck. The François family had the forward space all to themselves. The *Polaris* was stowed on board upside down and this made a good roof for their sleeping quarters; we towed the *Aldebaran*. The space aft of the mast we used for our own sleeping quarters; it was big enough to hold two camp cots set thwartships. We had only these two cots, but as Lion and I were never asleep at the same time one was enough for us, and the Doctor had the other. The small space aft was our kitchen; we had two Juwel oil lamps, on which all our cooking was done. Lion and I kept watch-and-watch proper deep–sea fashion. François was crew, and stood by to help, but the only time his services were required was when we came to or left shore, which wasn't often. The Doctor was cook; we had meals at the change of the watches: breakfast at eight, dinner at noon, tea at six. We lived for the most part on hominy, hard–tack, erbswurst, salt pork; with

The *Jupiter* under Full Canvas

chocolate and tea to drink; a diet that did not seem to inspire the François family with much enthusiasm, though they liked the erbswurst well enough. They called it "beans" which wasn't a bad guess.

Lion finished out the first watch, and when I relieved him at noon our favourable breeze had nearly died away and we were drifting slowly along the shore with scarcely more than steerage way. This part of the lake shore is well wooded, almost to the water's edge; the shores are low gravel hills. The weather turned dull and threatening, the wind sprang up from the N. N. E. and we fell off from our course a good deal. We had never been unduly optimistic in our expectations of the *Jupiter's* sailing qualities, but she made even more leeway than we expected. I don't think we ever made better than within seven and a half points of the wind.

It turned cold and more and more unpleasant, and the wind freshened considerably. I took the 6–8 dog watch; at the end of that time we were far out in the lake, much to the consternation of the François family, who evidently thought our methods of navigation quite wrong. For the Indians simply poke along the shore; if they have a wind dead aft they sail, if it is calm they row, and if a head wind they tie up till it changes, and tie up for the night in any event. This method of navigation is all right when there is a crew of eight or ten men, but with our desperately short–handed vessel we wanted to keep clear of the land; it was nothing but possible danger to us; we hadn't even an anchor. So although we were getting rather badly off our course we had really no choice but to stand on as we were.

The continual daylight was of the greatest advantage to us now; without that our voyage would have taken much longer.

It was cold, raw, and threatening when Lion took the tiller at eight; the dark clouds over the now distant spruce forests looked gloomy enough and matters got worse as the night came on. I tried to sleep my watch below but I was chilled through and it took a long time to get warm; the sea got up and the old *Jupiter* began to labour heavily. At midnight I turned out to relieve Lion; we had been blown far off our course and the high hills on the south side of the lake were visible. Close to the north of us was a low, rocky island; no islands are shown here on any map of the lake and it is quite possible that we were the first to see this one, particularly when the stick–to–the–shore methods of navigation in the north are considered. But it was rather curious to come across such an island so unexpectedly in the weird midnight dusk of a wild night; especially to Lion, accustomed as he was to working with accurate charts.

The weather got thicker and the land was soon hidden by mist. The sea ran higher and higher, the old *Jupiter* wallowed along through it, her timbers groaning and her sides swelling in and out like an accordion as the strain was thrown on and off the weather shrouds. I expected something to carry away any moment. Lion, who had turned in but who was sleeping no more than I had done, said afterwards that he expected the whole side of the boat to give way. What François thought we never knew, probably his fears of a sudden end were partly compensated for by the feeling that his ideas were now shown to be right; and that our methods, carried out in spite of his protests, had been proved quite crazy.

About 2 A.M. I sighted two more smaller islands to the north of us. The mist cleared off a little and I could see the high Great Bear mountains again, and out in the lake, some eight or ten miles ahead, a long line of spruce trees. I thought this was another island close to the mainland, and as we could just about fetch it on our present course I hoped to find shelter under its lee. By 4 A.M. we were close to its southern point; it proved to be a long hook–shaped point enclosing a bay several miles long and about a mile and a half wide, a most welcome harbour to us at the time when we most needed it. In fact we had struck the only place on this coast that was safe from a north wind.

It was delightful to pass into the quiet water behind the point; the wind moderated and it turned out a lovely day. We stood across this bay, the shore was lined with dense spruce forest coming down to a sandy beach; these were the biggest spruce trees we had seen since leaving the Slave River. Behind the dark spruce forest rose the high Great Bear hills with large areas of brilliant green; they were too far away to determine exactly what trees or bushes these were, but they looked very cheerful in contrast to the sombre spruce.

The water was so shallow we could not get within two hundred yards of the shore; it was a sandy bottom with rocks here and there. We landed in the *Aldebaran* and got a couple of big stones with which to anchor the *Jupiter*, then we all went ashore and had a comfortable breakfast on that beautiful beach with the big spruce trees above us and the placid bay in front, a pleasant change indeed after the last twelve hours. Especially did the François family seem to appreciate solid ground again, but they never were demonstrative under any circumstances.

We took things easy today; the wind died down, and by evening it was quite calm; so, after another comfortable meal ashore, we embarked again

A Welcome Harbour

and with much labour rowed the *Jupiter* across the bay and anchored off the point ready for the first fair wind. We could work only two sweeps; it took us more than an hour to make that mile and a half. I was on watch 8–12 and spent most of the time ashore on the point, hoping to get some duck. There were large numbers of small gulls; they highly resented our presence and kept swooping and screaming around the boat. Lion tried fishing on his watch, but didn't get anything.

The next day was dull and rainy; we all had our breakfast ashore, then the wind sprang up from the southwest, so we stood out in the lake again heading for Gros Cap. This day and the succeeding, July 18th and 19th, were dull, cold, and rainy, often with very heavy showers. The winds were light and variable, and though we made some progress it was by a very erratic course. About half-way across the lake we ran into a field of ice and had to make a detour to the west to avoid it. We went so far west, in fact, as to see the two islands again that we had sighted on the night of the 16th. We got to the north of this ice and shaped our course for the cape again. Early in the morning of July 19th we passed some reef just showing above water; we were then off the entrance of Richardson Bay and some fifteen or eighteen miles from it. There was more ice here also. We felt the cold a great deal during these two days, though the thermometer was four to eight degrees above the freezing point with very little variation; but it rained so much and so often that everything was more or less damp. Meal hours were the only bright spots in the day, and a good bowl full of steaming hot erbswurst the summit of happiness. We had brought a small amount of good brandy and rum "for emergency"; the brandy didn't seem to help us much, but the rum, an excellently good Jamaica rum put up by the Hudson Bay Company, certainly warmed the blood in our veins. This is one of the few times in my experience that I ever felt any direct and positive good from the use of alcoholic stimulants. Personally my bias is against them as such, and as a matter of fact we never had another real occasion for their use all the time we were in the North. But so much has been written against them, often indiscriminating nonsense, that it is a pity to miss an opportunity of recording the undoubted good they did on this occasion.

It was not till the morning of July 20th that we made Gros Cap under a light fair wind. As we passed a point we saw a York boat moored to the beach and knew this could only be the Hodgson outfit on their way back to Fort Norman. We landed at their camp and had the pleasure of meeting Mr. Hodgson, a fine-looking old man, a well-known old-timer in the North. They had left the Dease River some three weeks previously and had made their way slowly along the south shore of Dease Bay which they reported to be now free from ice. A fair wind was being wasted so we could not spend much time with them. Before we left Mr. Hodgson kindly gave us permission to use the shack he had built at Dease River, if we wished.

We said goodbye to them and continued our voyage, but we had gone only a few miles when the wind dropped, then changed to the north bringing up dense fog. We had to stand back to the land again; fortunately the wind was no more than a gentle breeze and we felt our way back carefully through the fog, coming to anchor behind a slight point that gave us a certain amount of protection. We had no idea how far we really were from the point on which the Hodgson party were camped.

The Doctor and I started out in the *Aldebaran* to follow the shore farther to the north and see if there was any better harbour near; we were quite unprotected from any east to northeast wind, and for anchors we had only rocks. It was quite interesting and mysterious to follow an unknown coast line through a dense fog, but it soon trended so much to the west that we gave up hope of finding any harbour in this direction. We landed and climbed a bare gravelly hill. It was pleasant to get ashore again, everything was strange and interesting. We had suddenly found ourselves in quite a foreign country; for the shrubs and plants were completely different from those on the south side of the lake either in species or in character. The spruce trees were very stunted and scattered, there were none near the lake at all. There were lots of flowers, all strange to us, some of them gave quite a heavy smell of clover to the air, bringing up vivid recollections of a country very different from this.

While we were poking around on shore the fog lifted long enough to enable us to see over the lake for a few miles to the north and west of us; it was covered with ice as far as we could see in that short distance; no open water was visible at all. We did not investigate to find whether this was fixed ice or floes, and at the time we were doubtful which it was. From later experience we know they must have been floes. There was no use looking for a harbour in this direction so we went back to the canoe; and while returning to our ship the fog came down thicker than ever and we nearly rammed the *Jupiter* before we saw her.

We tried south and found the land trended to the west in that direction also. Then the fog lifted and we could see that we were near the point of a wide shallow bay, evidently five or six miles at least from the place where we had seen the Hodgson outfit. We returned to the *Jupiter* and decided to stay where we were anchored and await the pleasure of the wind.

We landed and put in the time hunting ptarmigan, our first meeting with these birds which proved such good friends to us later on. The fog hung around the lake, but it was nearly always clear and bright ashore; the open hilly country with a few scattered spruces looked very inviting after the monotony of thick unbroken forests.

We made camp ashore that night, Lion and myself keeping our watches as usual, but with the unusual comfort of a cheerful fire.

Next day, July 21st, was still foggy and the wind still very light, but it had come around from the south. This was all in our favour, so we made a start and stood across the lake on a course for Cape McDonnell, the nearest point on the other side. We made slight headway till afternoon when a steady breeze sprang up from the southeast and then we made good progress. We left most of our troubles behind us at Gros Cap; thereafter the voyage was nothing but a pleasure trip.

The first land we saw was the high mountains far inland on the north shore of the lake. This was at 10 o'clock that night ; the sun was then nearing the horizon, and with beautiful effects it slowly dipped behind them. Cape McDonnell was abeam at 11P.M., but we had made so much leeway that we were now far to the north of it. This is low–lying land and we could barely make it out. Next morning we were still on the same tack with the Narrakazzae Islands ahead of us. But our leeway was so great that we passed clear to the north of them.

It was a perfect pleasure to be on board the *Jupiter* that day, the weather was lovely beyond description, and we had the interest of land visible in all directions ahead. We kept on slowly all day, doing the best we could with the *Jupiter* close hauled; by five in the afternoon we brought up at a low peninsula on the north shore of Dease Bay, about thirty miles from the Dease River.

We could do no more till we had favourable winds; it was annoying to be hung up almost in sight of our port, and time so precious, but patience was our only resource. We had a snug little harbour, a comfortable camp ashore, and plenty to interest us in the novel character of our surroundings.

It was the afternoon of the next day, July 23d, before a favourable breeze came; we started again but only to be driven back by a sudden change of wind. Lion and I were standing our watches all the time and we made another start on July 24th, at 3 A.M. Lion got the ship under way singlehanded. I came on watch at 4 A.M. that morning; we were then running before a gentle west wind and what we took to be Big Island was some fifteen miles ahead of us. Everything looked promising for a clear run, when the wind suddenly changed to southeast and a dense fog came up. All I could do was to keep the *Jupiter* as close hauled as possible, though we were fast being driven into some kind of a bay. I expected to find ourselves ashore any minute, as this part of the lake is full of islands, shoals, and points. Luckily for us the wind was only a gentle one, and after an hour of anxiety it suddenly changed again to the southwest and soon increased to a good fresh breeze. I could now only make a wild guess where Big Island lay, but I made the best guess I could on the compass and headed for that. The wind freshened and the *Jupiter* showed a capacity for speed quite unsuspected.

The fog was soon blown away; at 8 A.M. we could see Big Island about six miles away. My guess had been a good one, and we were heading direct for the north channel between Big Island and the shore. Behind us we could see, when the fog lifted, a long point that we must have just missed by I don't know how slight a margin.

The wind still freshened and the *Jupiter* fairly boomed along in front of it; certainly she could travel when she got the right conditions. At a point to the west of the site of old Fort Confidence we saw a camp of

Indians; they were in the wildest state of excitement at our appearance and ran along the shore shouting and waving their arms, but we never stopped to talk to them and the fastest runner was soon left far behind.

On we went past old Fort Confidence, now only a bare grassy space surrounded by spruce, with four chimneys alone left standing. Farther on we

Remains of Old Fort Confidence

saw a small log shack which we knew must be Hornby and Melville's house; it looked like a diminutive chalet among the rocks and spruce. We sped across the beautiful bay behind Big Island with its rocky spruce–covered shores and rock islets; the nearer we got to the end of our journey the faster the *Jupiter* went; we entered the Dease River flying, regardless of possible sand–bars or shoals. On and on up the river we went in triumph; in spite of banks and bends the wind held fair and strong; the *Jupiter* never stopped in her wild career till she ran hard and fast aground on a gravel bar in mid–stream, just below the first rapid.

Our voyage was over; whatever hard thoughts of the *Jupiter* we may have had at times, they were all forgotten now in her brilliant performance at the finish. Indeed we had good reason to be thankful, our voyage had taken eight days, but absolutely dependent as we were on fair winds it

We Arrive at the Dease River

might well have taken a month. We had had a pretty rough experience at first with stormy weather; with ice, rain, and fog. In stormy weather we had found a good harbour when most needed, the other incidents were at worst mere discomforts. The latter part of the voyage had been delightful in every way; the triumphant finish made amends for all the trials of the past.

We were aground a few hundred yards below the first rapid. Lion and I paddled the rest of the way and landed just below them. The river here makes a horseshoe bend enclosing a long level point thickly covered with spruce trees. On this point Hodgson had built his house. It was a rude, poorly built log shack about 18 ft. long by 16 ft. wide, so poor a thing that we decided at once to build another; and after getting the Doctor and looking over some possible sites on both sides of the river, we finally decided

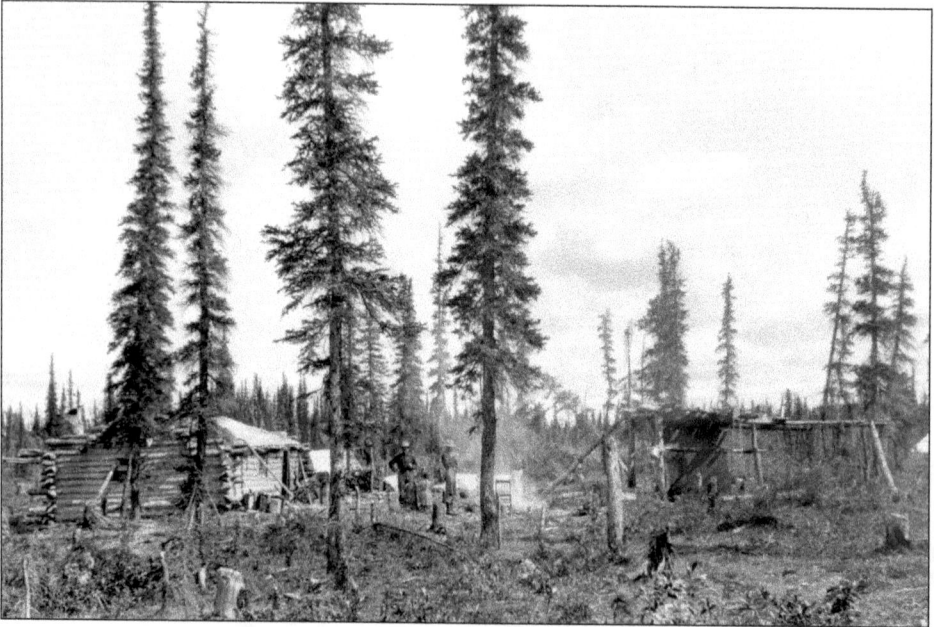

Hodgson's Shack

to build farther out on the point, about a hundred yards from Hodgson's house.

It was a charmingly situated spot, the river curved around the point, with a high sandy cut bank on the opposite side which gave us good protection from the north. We were delighted to find the spruce so much bigger and better adapted for house building than we had ever hoped in our most optimistic moments. "Hodgson's Point" we christened the place; its only drawback arose from its very good qualities; we were so snugly sheltered that we had no view.

Our first business was to get the *Jupiter* unloaded; when we got back to her she was already afloat, the strong southwest wind had backed up the water in the river; so we brought her alongside, just below the rapid. Hodgson's house was only a couple of hundred yards from here; we began to

Unloading the *Jupiter*

unload at once, packing the stuff to the shack, which we intended to use as a storehouse, and very useful it proved.

The Indians we had passed appeared on the scene just as we had finished dinner; a grand tea-drinking and smoking party took place at once with the François family; we wondered what kind of an account they gave the others of their voyage across the lake. These Indians were rather a pleasant-looking set; we tried to get them to come next day and help us with the *Jupiter*, which we intended to haul up ashore.

By five that afternoon the *Jupiter* was empty and everything stowed

safely away in the shack. We put up two of the tents; Lion and the Doctor
took one and the François family the other; I slept beneath a brush shelter
left by the Hodgsons, a sleep that lasted from nine that night till 10 A.M.
next day!

Our plans for the season had been carefully prepared and we began

Preparing for our journey to the Coppermine

at once to carry them out. Lion was to remain here and build our winter
quarters while the Doctor and I were to make a preliminary journey to
the Coppermine. The next three days were busy ones for all: Lion first
took charge of the job of hauling out the *Jupiter;* I had to develop all the
pictures we had taken to date, then the task of getting together our outfit
and supplies for the journey the Doctor and I were to make up the Dease
River and across the Barrens.

By the evening of July 27th everything was ready for a start; we had
a most carefully chosen outfit for our particular purpose and a supply of

provisions calculated to last fifty days. In the meantime Lion, with the assistance of the Indians, had hauled the *Jupiter* high and dry to her last harbour, a safe and pleasant resting place on a willow–covered bank of the river.

And there she is to this day, her purpose served and her work well done.

The *Jupiter's* Last Harbour

IV

## THE VOYAGE OF THE "POLARIS"—OUR FIRST JOURNEY TO THE COPPERMINE

N OW came the really serious part of the journey; all our work hitherto had been of a preliminary character leading up to this main object. We intended to ascend the Dease River as far as possible, to cross the divide between its head–waters and the Dismal Lakes, and to descend by the Kendall River to the Coppermine.

Except a copy of a sketch map we obtained from the Canadian Geological Survey, we had no maps, for no really accurate map of this part of the country has ever been made. The sketch map we had purported to be a paced survey of the country between Bear Lake and the Coppermine River. But there were such evident discrepancies between this map and the accounts of those explorers who have been over this country before us that we placed very little dependence on it.

Our chief guide was Hanbury's book. In July and August, 1902, he crossed from the Arctic coast to Bear Lake by way of the Coppermine and Kendall rivers, the Dismal Lakes and the Dease River. It was the most difficult part of all his extraordinary journey and he describes it in detail; we found his account most helpful, accurate, and reliable; only in one place is there a small omission, perfectly excusable.

But it must be remembered that Hanbury was coming down stream, and for him all roads led to Bear Lake, while ours was the more difficult task of ascending, with possible roads at the latter part branching out in all directions.

The Site of our House

At 10 A.M. on Friday, July 28th, we started. The *Polaris* was loaded up; we took the last photographs of each other at the beginning of our respective tasks, the last goodbyes and good wishes exchanged, and we began our voyage.

The Dease River is a broad shallow stream, usually about 130 yards

The *Polaris* and her Crew

wide, though some of the rapids are only ten or fifteen yards across. When we started, the water in the river was at its very lowest; generally speaking it is not continuously swift but a series of shallow rapids, with the water running over gravel and small boulders; and quiet stretches intervening, where the water flowed deep and sluggish between high sandy banks, and where the paddles could be used to advantage.

We had to wade the canoe up the rapids; the water was seldom so swift that it required both of us at that job and one could handle the boat much better by himself. I usually waded the boat up, the Doctor stood by to help or lightened her by carrying some of the stuff.

It was all mercilessly hard work, harder really than even the tracking, but there was constant variety and intermissions of rest that we never knew on the tracking line. And we could always indulge in hopes in a way quite impossible when at the latter task; we could always hope that the next rapid might be the last, or at least so deep that it wouldn't be necessary to lighten the canoe; or that the next stretch of quiet water might last several miles. It is true these hopes were seldom justified, but enough good streaks came along with the bad to keep our spirits up.

Then we had good grub and good equipment, our camps were always comfortable, and we took trouble in preparing our meals. We had learned the necessity of taking good care of our bodies; they were mere machines for the conversion of heat into energy and required the careful attention necessary for every high class machine.

Altogether this journey would have been one of sheer pleasure if it had not been for the shortness of the season which left us so little time in which to do so much work. We couldn't afford to follow by ways inviting exploration, or to move at our convenience as regarded the weather; wet or dry rain or shine, we had to shove on.

This first day we had lunch about three miles above Hodgson's Point on a high sandy bank; there was an Indian grave there, the palisades looked quite fresh, and we learned afterwards that it was the grave of François's father.

While at lunch we saw a flock of young geese gravely swimming down the river in a long line; they looked at our canoe awhile, then decided to turn back. We overtook them later on and murdered four of them with a little .22 pistol. But our needs were imperative, and those young geese gave us some of the most delicious stews I ever tasted. We struggled on all afternoon; it was bright and warm, the river was a succession of shallow rapids with quiet stretches of deep water between, and we were well pleased with the progress we had made.

We camped at the head of an island, at a place where it was in any case necessary to portage all our stuff. It was an ideal camping spot, and we sat down to our supper of stewed goose that night in a state of complete content. This was No. I Camp, a place that became well known to us later on.

The Cañon

Next morning we made an early start, but things didn't come so easy as the day advanced. The river was very swift and very shallow, spread over a wide gravelly bed. We had to make constant portages of all our stuff, and in some places the river was so spread out and divided that there was scarcely enough water to float the empty canoe. At one shallow rapid I knocked a hole in her; we had to stop at once and repair it; I had brought some thin sheet brass for this purpose. We camped that night just above a limestone gorge through which the river cuts; the scenery is very picturesque, but somewhat depressing from its very ruggedness. We recognised at once a large pillar of rock standing in the river described by Simpson, whose men called it the "Old Man of Hoy." A hard day's work had brought us only four or five miles from our No. 1 Camp. This was not encouraging, and

The Cañon—Old Man of Hoy to Right

there was a conspicuous lack of enthusiasm in our attitude towards the Dease River that night, till after we had eaten and been cheered by a good supper. I climbed to the top of a hill behind our camp hoping to see some caribou. I saw no game but the view at this part was rather fine. A range of bare hills lay to the northeast, one high peak detached from the others was not far from the river; this had been visible since leaving the Indian grave.

We had a long portage to make next morning, and then, after surmounting a series of rapids, we came to a fine quiet stretch of water.

We decided to camp near the big hill and have a look over the country. According to our sketch map we were now only about fifteen miles from the Dismal Lakes. We didn't exactly believe this, but certainly the map had proved accurate enough as far as we had come, and we fervently hoped it might prove correct in other respects as well. It was in just these other

Observation Hill

parts that we needed guidance most and where the map failed to check up with Hanbury's and Simpson's accounts of the country.

We called this hill Observation Hill, but learned afterwards that the Indians call it Rabbitskin Mountain.

We got a magnificent view from the top; to the southwest the Dease River ran winding among spruce–covered hills, beyond lay Bear Lake; we could even see the Narrakazzae Islands and a high point on the north shore that had been a great landmark to us during the last two days of our voyage on the *Jupiter*. To the north the scenery was very different, a wilderness of sandy plain and sparse stretches of spruce, countless lakes and green marshes; far to the north, blue in the distance, lay a range of low mountains. It was quite impossible to trace the Dease River in this confusion of small lakes, and evidently we were still a long way from the Dismal Lakes; we concluded unwillingly that they must lie beyond those distant mountains in the northeast.

We walked farther along this high range of hills; it had been very warm on the river, but it was pleasantly cool here. The blue berries were ripe and grew in prodigious quantities on absurdly small bushes, little things not more than a couple of inches high, but loaded with big juicy berries. But the black flies were fierce, they assailed us in swarms; it is harder to defend oneself against them than against mosquitoes. We went back to our camp in good spirits; our disappointment at failing to see anything to indicate the proximity of the Dismal Lakes did not last long; we had not really expected to, and it looked as though we would have better conditions on the river next day.

And so indeed it proved, we made a good early start, shoving off in a pouring rain which lasted most of the day. But we had good going, with an uninterrupted ten–mile paddle for a start; it was a perfect joy to make progress like this after our late incessant wading and portaging, and we didn't mind the rain so long as we had such good conditions for canoeing.

The river was really very pretty although somewhat monotonous. It curved between high banks of clay or sand, level on top and covered with a thick growth of small spruce trees and usually with thick willows coming close to the water's edge. It was deep and sluggish, with a scarcely perceptible current.

All along the river animal life abounded. There were ptarmigan ashore and muskrats in the water; the latter paid very little attention to us but went about their business quite undisturbed. Every few hundred yards we would pass a mother duck with a small fleet of little ones. The old birds would come fluttering right to the canoe in heroic efforts to distract our attention from their young ones; it was quite pathetic to see how anxious they were to sacrifice themselves to give their flocks a chance to escape. But we never molested them though we shot some more young geese today.

The rain lasted all morning; during the latter part of the morning and

Wading

all afternoon we were struggling against swift and shallow water. We camped that night at what had evidently been the camp of one of the early explorers, probably it was one of Simpson's camps. There was even a pile of firewood cut ready for use; it was much weathered but still sound enough to make a good blaze. Little did the chopper of this wood think who would finally use it, or at what late date.

Our struggle against swift and shallow water continued next day; it was a case of wading nearly all the time at this part of the river. We had some dark and gloomy looking hills ahead of us all day; late in the afternoon we came to a place where the river made a sudden bend and we soon saw that it cut through a high rocky dyke that extends for many miles across the plain. It is a notable landmark, we called it "Notman Dyke." We had

Notman Dyke

to make a complete portage of our stuff for several hundred yards here and we camped at the end of the portage.

The day closed in dark and gloomy with a cold north wind. The view from the top of the dyke was inexpressibly wild and menacing; rugged barren hills and dark stagnant swamps were around us, and a threatening wind-swept sky above. It was a relief to turn into our snug little tent and forget all our troubles in sleep.

We still had swift and shallow water all next morning; then we came to a sandy plain through which the river wound in great loops. It was deep enough to use our paddles and we made fair progress, but so crooked was the river that sometimes after paddling a mile or more we would find that we had really gained only a few hundred yards. The river had shrunk very

Hanbury's Kopje

much in size now; we passed many small tributaries and one large one today.

At last after a hard day's work we were gladdened by the sight of a high sandy hill, its peculiar flat top covered with a thick growth of small spruce. For the last two days we had been looking out for this; it was the "kopje–shaped hill" described by Hanbury, and it was near this hill that he had struck the main Dease River on his way from the Dismal Lake.

About a mile below it we passed a tiny stream coming in from the northwest; the Doctor hailed this at once as Hanbury's "Sandy Creek." I wasn't quite so sure about it; the impression given by the book is that Sandy Creek comes in from the northeast, and this stream seemed too minute to answer Hanbury's description. We camped that night below "Hanbury's Kopje," as we called the hill, determined to investigate before we went any

Camp at Hanbury's Kopje

farther. It rained heavily all that night, our camp was right out on a flat sandy beach, but there was plenty of spruce for fire and bedding.

Next day it was still raining, an incessant light rain. We tried the right, the main branch of the river, first; we could paddle a short distance above the kopje; what a delight it was to paddle an unloaded canoe again! We soon came to stony gravelly country, the river was full of rapids, so we took to the shore and walked over these undulating gravel hills for several miles more.

But the farther we went the more evident it became that this could not have been Hanbury's route; we returned to our camp and had lunch, then started to explore the other little creek. It was a mere rivulet flowing in a wide sandy channel, with a line of gravel hills on one side and the sandy plain to the south on the other. We landed a mile or so from the mouth and struck across to a high range of hills about four miles to the north.

We called this Granite Ridge; it rose about five hundred feet above the river and commanded a very extensive view, but the weather was too thick to see anything that definitely indicated the location of the Dismal Lakes, so we returned to our canoe and paddled back to camp. We had not ascertained with any certainty that this little stream was Hanbury's route, but although it was so small it was navigable to a quite unsuspected degree and we decided to try it. So we went back to our camp on the sandy beach and made a good early start next day.

We found it was one thing taking an empty canoe up that little stream and another when the canoe was loaded with nearly five hundred pounds of stuff. It was desperately discouraging work, incessant hauling, unloading, packing, reloading, and hauling, and so on *da capo*. It took us most of the morning to reach the place where we had been with the empty canoe the day before. A short distance beyond this the stream branched, making our perplexity worse than ever.

We tried the left hand branch first; a small sluggish stream, so narrow and with bends so sharp that it was almost a matter of difficulty to get our eighteen foot *Polaris* around some of them. And we were getting so far to the west that the whole business looked very doubtful; at last we decided to make an attempt to reach the Dismal Lakes on foot and see where they actually lay. We made camp at the little forks on the creek, "Junction Camp" we called it, a charming spot and a snug camp it was.

Next day we made an early start in our heroic determination to reach the Dismal Lakes on foot. We struck across to Granite Ridge again intending to follow that along, but the weather came up thick and rainy, cold and raw, even a little snow fell; we could not see anything under these circumstances, so we waited awhile under a high rocky ledge hoping it would clear off. While waiting there, cursing the country, cursing the weather, cursing Hanbury and his descriptions (who certainly didn't deserve it), a big bull caribou came trotting along right towards us. It never saw or suspected us; we simply let it come to within easy range and then killed it. This was the first caribou we had seen, and we were surprised to find it so big. As a matter of fact it happened to be an unusually large one; we got only one other afterwards that ever approached it in size. The incident was a welcome diversion from the comminatory service we were holding and the caribou meat an even more welcome addition to our grub supply. We skinned it and cut it up, piled the meat on a high rock, and threw the hide over it.

Then the weather cleared up and we continued our journey along the crest of the ridge for several miles. At last we got right to the end of it and were confronted by a plain, in a depression of which lay a round, fairly large lake. Beyond this again was the range of mountains, still eight or ten miles away. We knew that the Dismal Lakes must lie beyond these somewhere, but there was certainly no prospect of walking there in one day!

From the end of the ridge we could plainly trace the course of the Dease River. It came out of a long lake to the east of us, and from here it looked like a promising route.

For we were now completely at loss to know where Hanbury had been, this at least seemed to offer some kind of a route and we decided to try it.

We went back to our caribou, then came a long hard walk back to camp, loaded with all the meat we could carry. We revelled in fresh meat that night; with the constant hard work and the wet cold climate, our appetites had grown enormous.

We returned to the main river next morning, and put in a long hard day ascending that. It got smaller and smaller as we went on, but in spite of constant packing and portaging we made fair progress, and evening found us among some steep gravel hills we had seen from Granite Ridge. This part of the country was unspeakably savage, ground up by old–time glaciers into barren gravel moraine. We camped on the shore of a little round lake through which the river passed, at the foot of one of the most conspicuous of these gravel hills.

I had a regular disaster the next morning, a disastrous opening to what seemed to be a disastrous day. I had put my boots to dry in front of the fire and while we were having breakfast one of them rolled in unnoticed and was completely spoiled. This was a bad business for me as I had only the one pair; after that I was reduced to wearing the heavy cumbersome wader–shoes all the time. My feet were always soaking wet, and worse than this the wader–shoes were most clumsy and tiring to walk in.

Things got worse and worse with us today. By noon we had not made more than a mile from our camp in a straight line, so difficult had the river been. Then it spread out over a wide gravelly bed, merely a trickle showing here and there; further progress by water was quite impossible. We went ahead on foot to see what conditions were like; the river contracted again, but it was hopeless for navigation.

We were still about five miles from the lake; this part of the journey would require a complete portage of our stuff and would take several days at least to get to the lake. And even once on the lake we were still a long way from the Dismal Lakes. We had come much farther to the east than we had suspected when looking over this valley from the Granite Ridge. From a hill above the south end of Long Lake, as we then called it, we could see nothing but rugged hills to the north. The season was getting so late that we might despair of ever getting there this summer by a route so palpably difficult. We discussed the subject in all its bearings, finally deciding not to waste our time and strength by persevering in an enterprise so doubtful, but to turn back to the little creek we had already attempted once, to follow it up as far as we could, and to stake the whole success of our journey on finding a way to the Dismal Lakes by that route.

We camped early that day at the farthest point to which we had brought the canoe; our camp was on a high gravel point overlooking the stony river bed, a waste of gravel a quarter of a mile wide with the water trickling over it in small streams. The weather had been lowering with frequent showers all morning; now it came on a savage storm with high winds and fine driving rain. We called this place "Camp Despair," and desperate indeed did our chances of success seem when we turned in that wild and stormy night.

"Penury, inertness and grimace . . . . . . . . were the land's portion"

Next morning broke fair and cold; the rain had stopped and the wind had ceased, our wet tent was frozen stiff. We broke camp, packing our stuff a quarter of a mile down the river, saving thereby several portages, then began the weary job of retracing the journey we had made so confidently. But we made much better time going down stream, and the rains had swollen the river a little. We camped at junction Camp again that evening.

For the second time we shoved our way slowly up the little sandy creek. About two miles above the point at which we had turned back before we came on some features that made us think that this, after all, might have been Hanbury's route.

And so indeed it proved; a mile or so farther on we reached a place that answered exactly to Hanbury's description, and not only to Hanbury's but Simpson's as well. We had always supposed from reading their books that they had used different routes, now we could see it had been the same. Later on we had proof of this; we found the remains of camps with chopping of the same age and character as at some of the old camps we had seen on the lower river and had been able to attribute to Simpson; one of these camps was in a situation exactly as described by him. In fact, except on the lower part of the river where there were frequent signs of the Hornby and Hodgson parties, all the indications of any one having been there before were old and had been made by Simpson, Richardson, and Rae. We found no signs on our whole journey that could be attributed to Hanbury.

Then followed three days of constant effort. At one place we had to make a complete portage of all our stuff for more than two miles. But in all this hard work we had the encouragement of knowing that we were on the right road; we could work with a will, free from the doubts and perplexities of the past week, and worst of all, the haunting sense of possible failure.

The river got smaller and smaller; we passed from sandy plains to low rolling gravel hills, then to sandy hills again. The stream finally became a mere sluggish ditch, meandering with many turns through a narrow valley; the hills had closed in on us almost before we knew it. It was evident that we had followed the creek to its very source; before us lay an opening in the mountains that we thought might be the divide that Hanbury had crossed. At the nearest point to this we made camp; it was noon of the third day after leaving junction Camp.

We started out on foot to explore the country; the walking was good over gravelly hills that skirted a plain on which many small lakes were scattered; after travelling a few miles we came to a ridge overlooking a long narrow valley, with a series of small lakes in it. This was Hanbury's route beyond question, and at the end of this ridge we came in sight of a long tongue of water lying among the mountains. We had reached the Dismal Lakes at last.

While taking a rest before starting back to our camp, I was surprised

Along the Divide

to see a man walking around on the hills about a mile away from us. He disappeared just as I brought my glasses to bear on him, but the glimpse I got showed him to be an Eskimo. We could see some kind of a little camp on top of the hill and we headed for this, highly excited at the prospect of meeting these people whom we were so anxious to see. We made a cautious approach, fearing to frighten the man, and we were quite close to him before he noticed us. Whether it was merely the unexpectedness of it, or whether he had never seen any white men before I do not know, certainly he was very much frightened. We threw up our arms calling out "Teyma! Teyma!" about the only Eskimo word I knew. He did the same, his arms were fairly trembling with fright, and he kept repeating something over and over in a low moaning tone. However he was soon reassured and smiling.

Dismal Lake

He was a stoutly built man, about five feet four inches high. His hair hung straight and black behind, all the front part of it was cropped close to the skin. His face was open and intelligent, with rosy checks and a candidly engaging smile. He was dressed in caribou skins and sealskin boots, his general appearance was quite as clean as our own and a pleasant contrast to the dirty, sulky Indians we were used to. He had some spears, and a bow and arrows in a sealskin case lying on the ground; supported by four cross sticks was some kind of a roll of skin. We tried to talk to him asking if that was Teshierpi Lake. He nodded his head and repeated "Teshi–arping, Teshi–arping." I gave him a small piece of milk chocolate I had in my pocket; he put it in his mouth in a doubtful way. Presently an expression of intense delight passed over his wide face; it was very funny to watch.

We tried to explain by signs that we would like him to come back to our camp and help us to portage our stuff. He followed us willingly enough for awhile; whether our pace was too fast for him or whether he was afraid to trust himself with us so completely I do not know, but he lagged farther and farther behind; finally we saw him make a beeline back to his camp; he picked up the roll of skin, gathered up his bow and spears, and disappeared toward the lake.

We were not altogether sorry that he turned back; we had no fresh meat left and were doubtful how he would have liked the only food we could have given him. It was late when we got back to camp that evening; the nights were now beginning to get dark and I saw some of the brighter stars tonight for the first time since leaving Fort Smith.

It took two days to make the portage; the distance was six and a half miles, but we were helped out by the small lakes, which made up perhaps one–quarter the whole distance. We moved all our stuff along in short trips of about five hundred yards; it took three loads each and one trip with both of us to carry the canoe, so that we had to cover the ground seven times. But the walking was good and the little lakes gave us most welcome rests.

It was a joyful evening when we finally got all our stuff to a pleasant little camping spot at the extreme end of Dismal Lake. We celebrated the occasion that night by an extra good feed, with coffee and desiccated raspberries as a special treat. We had brought a few tins of coffee for great occasions; the desiccated raspberries were, alas! a great disappointment. But their juice was fine; we mixed it with a little brandy of which we had one bottle "for emergency" and it made a drink that we thought was perfectly delicious. We called this "Teshierpi Toddy"; it figured at festive occasions later on, and it was the only use we ever had for our desiccated raspberries. It was a wild and stormy night, but we were in good spirits; our little camp was snugness itself; we had accomplished the most difficult part of our journey, and we hoped for an easy road the rest of the way. Moreover, in spite of the time lost around Hanbury's Kopje, we were still only two days behind the estimate I had made more than six months ago when planning the expedition.

We took things easy the next morning; it was seven before we got up, our usual hour was five. It was still dull and threatening, but the wind had greatly moderated. It was a joy to load up the canoe and to put on her canvas covers, sure that we would have no occasion for the everlasting handling of stuff that had hitherto been necessary.

It was this part of the lakes that Simpson saw when he gave them their name; there never was a name better deserved so far as the western extremity is concerned, though it is unmerited as regards the rest of them. Anything more unspeakably dismal than the western end I never saw; the lake is shut in by high bare rocky hills; those to the north still had huge drifts of snow in places, probably these drifts never disappear. The hills on the southern side are sharp broken rocks unmitigated by any softening influence of plant life. Five or six miles from its western end the lake narrows to half a mile or so in width, a sombre sheet of water between threatening hills. Beyond these narrows the character of the lake changes; instead of the bare rocky hills there are gently rounded slopes covered with grass and heaths, wholly pleasant in aspect.

We paddled along the south shore of the lake; by noon it began to rain heavily and kept it up all day and all night. We got to the first narrows at 6 P.M.; we must have paddled at least twenty–seven miles today. A sandy bar stretches across the lake here; it was so shallow that we had to wade. The south shore is a series of small sandy cliffs, the north shore consists of low willow–covered stretches, and there is a low sandy island also covered with willows. I walked along the shore; it was strewn with the bones of caribou; bones of all ages scattered like driftwood along the beach. These narrows are a favourite crossing–place for the caribou; vast numbers must have perished here at various times. We entered what appeared to be a river, a narrow passage between steep gravel hills with a quite perceptible current. Along the hills we saw lines of stones set up on end, each one crowned with

Camp at the Narrows—Dismal Lake

a tuft of grass; this was the work of the Eskimos in their systematic warfare against the caribou.

We camped on a level mossy spot at the foot of one of these gravel hills; it was pouring rain, fire was out of the question, we had nothing but moss and heaths to make it of. We just put up our tent and turned in, thankful that we had plenty of hardtack and even yet some milk chocolate.

It was a lovely morning next day, clear and bright; everything looked fair and fresh in the unaccustomed sunlight. With a great deal of trouble we got together enough dry willow twigs to make some tea and boil a pot of erbswurst. We thought of Hanbury's remarks on the ease of making a fire at all times on the barrens, and wished we could have had him there that morning to show us.

Not till after breakfast, when I climbed the hill behind our camp to

Eskimo Kayak and Cache

have a look around, did we discover that there was some kind of an Eskimo camp not far from us. A kayak was carefully put away on some little stone trestles and loaded down with rocks to keep the wind from blowing it about, and various other small belongings, fur clothing, etc., lay around.

It was at this point that Hanbury met Eskimos on his journey through the Dismal Lakes, and evidently it is a favourite camping place with them, for there were traces of many old camps in every direction: small fireplaces, scraps of bone, and refuse of various kinds and of all ages.

But there was no sign of any one, and there was nothing about the camp to indicate how long it had been deserted. We feared that the Eskimo we had seen at the end of the lake had given the alarm, and that they had left this part of the country. We placed a few needles and files around, the camp, I put a piece of milk chocolate under the kayak so that they could identify their visitors, and then we continued our journey.

The middle one of the three lakes making up the Dismal Lakes is much shut in by mountains. Those on the north side rise abruptly in terraces to nearly a thousand feet above the lake. On the south side is a high rounded hill; it rises eight hundred feet and seen from the north and east it is quite regular in shape; we called it Teshierpi Mountain; it is a notable landmark. Though so much shut in by hills the middle lake is not gloomy, Teshierpi Mountain showed slopes of green coming right to the water's edge, even the sharp cliffs of the mountains on the north side were interspersed with cheerful grassy terraces. The second lake ends in a shallow sand–bar stretching right across the lake, then comes a tangle of small sandy willow covered islands. There is a difference of elevation quite perceptible; the lakes are connected by a short river flowing through a sandy channel. Quite a large stream comes in from the south at these narrows; we did not discover it till on our way back when we called it the Teshierpi River.

The last of the Dismal Lakes is the smallest and most beautiful of all, a delightful sheet of water about three miles long and a mile or less wide, with gently rising grassy shores and surrounded by mountains at just that distance at which they are most charming, close enough to give an impression of intimacy and protection, and far enough away to be free from any feeling of imprisonment. Like the other two lakes there is a shallow sandy bar extending right across its eastern end; a small river comes in from the southwest; Richardson mentions this stream.

We entered the Kendall River; it is considerably larger than I expected to find it. No doubt it had been swollen by the continual rains, although the water was brilliantly clear. We found small spruce again at the extreme end of the third lake and the Kendall River flows between banks covered with sparsely wooded small spruce. We camped that night about a mile below the lake expecting that a few hours' run next day would take us right to the Coppermine Mountains.

Our food supply was beginning to worry me; our consumption had been far in excess of what we had figured on. The caribou we had shot had helped us out, but we had been able to take only a small amount of the meat and that had soon spoiled in the continual wet and constant handling. We had got young geese, ptarmigan, and ducks from time to time along the river, but we couldn't afford the time for systematic hunting. The only fish we had obtained so far had been a few Arctic trout that we had shot in the shallow rapids of the upper river. We had tried trolling when passing through the Dismal Lakes, but here again we could not afford the time and had thrown our bait into the water only when we were passing some of the long gravel points, as Hanbury says this is the best place to catch the lake trout. A year later when crossing Great Bear Lake and fishing for our daily food we found that we could always catch plenty of lake trout trolling in the small bays.

Tonight we got out our net for the first time; however simple it seems in the abstract I soon found that to set a net properly requires, like any of the skilled trades, a high degree of manual skill; and besides this, an extraordinary command of one's temper. But I finally got that net set, after a fashion, and we hoped for some of those fabulous hauls of fish that we had heard of.

I turned out very early next morning; my first care was to examine the net and I found we had caught one white fish of about three pounds weight. And that wretched fish had tangled up the net worse than ever; by the time I got the fish extricated and the net ashore it looked like a stringy bunch of cotton waste.

I crossed the river in the canoe, throwing out a trolling bait as I did so and catching a fine lake trout of about five pounds.

A short distance above our camp, on the opposite shore of the river, I came on the remains of a quite extensive camp; this can be identified without any doubt as the remains of the provision station established by Simpson in the spring of the year 1838. The place seemed gloomy enough to me that early rainy morning thinking about our food supply and the unknown dangers of the Kendall River. The picture was taken under more

The Kendall River—Remains of Simpson's "Spring Provision Station"

favourable circumstances on our return voyage, when that stretch of the river and the mountains in the background looked lovely indeed.

We broke our camp in a pouring rain, and, making everything extra snug aboard the *Polaris*, we started our run down the Kendall River.

The river is much like the lower Dease River in character, but everything is on a larger scale, except the spruce trees along the shores, which are shorter and more stunted. About four miles from the lake we landed and climbed a high limestone hill not far from the river. Hanbury mentions this hill and the view obtained from the summit of the most easterly Dismal Lake. Today the country looked most forlorn and melancholy, with the more distant mountains hidden in mist and over everything the incessant rain.

The unexpected length of the river surprised us. Hanbury supposed

Cañon near the Mouth of the Kendall River

it could be run in a couple of hours; evidently we had a much higher stage of water than he had, and we ran quickly with delays at only one or two of the rapids.

We ran all morning; lunch time came and we had a feed of dry hard-tack sitting on the muddy bank of the river in the pouring rain wondering when we would get to the end of our run.

All afternoon we continued the descent; the constant strain of steering down the boulder–strewn rapids began to tell on me, the farther we went the worse they got. I can well remember my dismay later on in the afternoon, at a time when we certainly expected to have left the Kendall River far behind us, to find ourselves at the head of a long, dangerous rapid full of big boulders. It was a regular hill, and we went down that place like a toboggan slide.

Junction of the Kendall River with the Coppermine

At 5 P.M. we came to a small canyon through limestone cliffs and we knew that we were near the Coppermine River at last.

At the top of the canyon we landed to look ahead, the water was swift and at one place there was an ugly "set" against the jagged, perpendicular wall, but I thought that I could avoid this, and we started out. There is a swift rapid at the head of the canyon; at the bottom of this rapid we struck a big boulder and knocked a great hole in our canoe. The jar and momentary stoppage swung me off my course and we got carried into the worst part of the "set." Fortunately I had been able to get the canoe bow ahead again before we got there; we skimmed past that rocky wall with scarcely an inch of clearance; the Doctor did valiant work shoving off and we

The Coppermine River

came through the canyon in triumph, running the canoe ashore on the beach below just before she sunk.

We found ourselves on a level grassy point between the Coppermine and Kendall rivers; we made camp at once and had a good supper off the lake trout I had caught in the morning. The rain cleared off later and we even had a lovely streak of sunlight on the opposite shore of the Coppermine.

That river is spread over a wide gravelly bed here with low gravel islands and bars; a large river in reality at this point it seems even larger.

The water was clear, of a pale greenish colour, but not the perfect clearness of the waters we had been accustomed to of late.

It took me all the next morning to repair the *Polaris*; she had suffered severely on her voyage down the Kendall, and the last blow had been a

We Arrive at the Coppermine Mountains

cruel one. But she was a beautifully built boat, made of specially selected wood and by men who took a keen personal interest in their work. Good reason had we to be thankful for the trouble and care that her builders had taken with this boat, and by noon the *Polaris* was all right and tight again.

While I was working on the canoe the Doctor had the more difficult job of unravelling the net; by the time the boat was repaired the net was ready and we set it in an eddy at the juncture of the rivers, making a good job of it this time. It was there for nearly twenty–four hours, but we caught nothing. We spent the rest of the day hunting and prospecting.

On the morning of August 17th we started again; it was a lovely bright day and the conditions for canoeing on this part of the Coppermine River

Camp at the Coppermine River

were delightful indeed after our late experiences of rapids and rocks. We ran swiftly and smoothly past some high limestone cliffs along the river; we had walked over these the afternoon before. After passing them the river flows through a pleasant open valley with pretty little groves of small spruce scattered along the shores. Leaving the valley and passing among some high gravel hills, the river then cuts through the Coppermine Mountains, making a big sweep to the east as it does so. We camped on the west shore, right at this bend of the river, not far from a small creek marked on Franklin's map; the point at which we had beforehand decided to commence operations.

Our first job was to make a good camp; then we started out, the Doctor to make his first acquaintance with the copper-bearing rocks, I myself to

Coppermine Mountains—Characteristic Terraces

try and get a caribou, though I was beginning to despair of seeing any, as we had come through the best caribou country and had seen only one. The food supply was running short and causing us some real concern. But luck was with us today; at only a couple of miles from camp I saw a big bull caribou wandering along a level grassy depression among the mountains. I made a cautious approach and fired at about three hundred yards, not daring to risk frightening it by going closer. It reared up just like a startled horse and began to run around in a circle; I fired a couple of times without apparent effect, and began to be afraid it would start off in a straight line instead of keeping in a circle, when to my great relief my fourth shot brought it down. Poor brute, the first shot had just taken out one eye and the next two had grazed it in different places. The

Headwaters of Stoney Creek—North Branch

shortness of our food supply didn't leave me with much sympathy for the caribou at the time. It was a large bull, about the same size as the first one we got, but fatter and heavier. These first two were the biggest caribou we ever killed.

I returned to camp with all the meat I could pack; it was a joyful greeting and a good supper I could give the Doctor when he came back later on in the evening. He too had been fortunate among the rocks; our first day on the Coppermine was in every way a most pleasant one.

The next week was spent among those mountains making preliminary explorations in various directions. The results on the whole were very encouraging, and a full and proper description by Dr. Sandberg will be found in the Appendix.

Coppermine Mountains and River

The mountains are basalt, rising with terraced sides about 1000–1200 feet above the river. It is an interesting and not a difficult country to explore, the walking is nearly always good, and the terraces give one a choice of easy routes.

There is a fair quantity of spruce, but only in the immediate vicinity of the river; in some sheltered places it grows much larger than one might expect so far north. We saw some trees as large as seventeen inches in diameter two feet from the ground. But none of these attained more than twenty–five or thirty feet in height, the average height of the spruce along the bank would scarcely exceed twenty feet, and they are all of a twisted and rugged character.

We had a few fine days while we were at the Coppermine Mountains

The Road Home

but the weather was usually wet and cold. We would have liked to have continued our journey to the sea; the descent would have been easy enough but the return would have taken more time than we could afford. And it was necessary to turn homewards now; the short northern summer was over and there were many signs that winter would soon be on us.

Our stay at the Coppermine had really been a very pleasant one, the comparative rest and good living did us a lot of good, and we were both getting stronger and stouter every day.

A View down the Coppermine River

We moved our camp to a point six miles up the river to a point on the opposite side and made some cursory explorations to the east. The mountains were lower, but as far as we went they were composed of the same series of basalts and traps.

The last days we spent on the river were delightful, we had lovely weather; bright days, ineffably serene, the land lying bathed in strong sunlight. We had sharp frost at night and usually we woke to find a fuzzy lining of frost on the inside of our tent.

The ascent of the Coppermine to the Kendall River was easy. Then came the long struggle up the Kendall River which took two days of toil and fatigue, but we were now more expert at wading and more inured to hard work. I was well satisfied that it took no longer and that we had got up a difficult stretch of our journey in safety. Hanbury's description of his feelings in safely accomplishing this part of his journey so well described my own that I shall borrow that intrepid traveller's words.

Like ourselves they had taken two days to ascend the river, but they evidently had a much lower stage of water and they had five men to their canoe so that they could use the tracking line at places where we could not. When they had finally reached the Dismal Lake, Hanbury says: "We coiled and stowed away the tracking lines and resorted to paddles. It was indeed a relief to get out on the open and quiet water of a lake. The ascent of a dangerous river, or rather I should say a river where continued caution is absolutely necessary to prevent an accident, is apt to get on the nerves. Every day the attention is strained and every night you are obliged to camp close to the thunder and swish of the rough, heavy, and rapid water which you know you will have to tackle the following morning. Very old hands may not experience these feelings, and very young hands are ignorant of the real danger that awaits them in the event of the bad canoe accident. For my own part I have not yet got hardened to risks which from ample experience I know to be serious."

It was the descent of a river that had got on my nerves, mostly, I think, on account of the difficulties I foresaw we would have on our return. But when it came to the point these difficulties smoothed themselves out and we made the trip in less time and more pleasantly than I had dared to hope.

Since we had descended the river only some ten days had passed, yet in that short time a complete change had taken place in the appearance of its shores. On our journey down it had been the height of summer, the mosses on the hills, the willows on the banks, the sedges in the river all wore vivid shades of green. Now the whole colour scheme had changed, scarcely a green leaf was to be seen. The mosses were every conceivable shade of red, wonderfully brilliant; the willows were a uniform bright yellow; the dwarf birches yellow also but paler and less brilliant; the sedges growing along the shore in the water were all mauve and cerise. It was the most beautiful display of colour in flora conceivable.

We spent a few days on the north side of the middle Dismal Lake attempting to investigate the country to the north. "Glacier Cove" we called our camp; it was right under a steep gravel hill, with hills to protect us on either side. A level space lay between us and the lake, a regular floor of fine hard gravel covered with a carpet of short thick moss, of all beautiful shades of red. It was a snug camp, and among the hills behind us there were lots of dead spruce for firewood. This is the only place where spruce occurs on the Dismal Lakes, except at the extreme eastern and western ends.

The weather came up thick and stormy; the mountains were hidden in mist and it was impossible to carry out our plans. We did get to the top of the range on one occasion, but it came up a dense mist and we had to find our way back to camp by compass.

Then came severe gales from the northwest; lucky indeed we had so snugly sheltered a spot. For two days we were completely stormbound, but we had plenty of fire-wood and plenty to eat, for we had shot another caribou coming up the Kendall River just as we had come to the end of our fresh meat. Eating was our chief amusement during the time we spent at Glacier Cove; the amount of food we could put away was perfectly amazing. All the time of our stay at the Coppermine we had lived for the most part on caribou, keeping our other provisions for the journey home; now we revelled in a varied and luxurious diet. We even put out our net again, but didn't catch anything in it, though we got some very nice lake trout trolling.

At last we gave up the idea of doing any more exploring this season The weather was turning cold rapidly and we had to get over the divide before lakes and rivers froze.

We made a very early start on the morning of September 1st; we broke camp and were under way at 6 A.M. It was a mild, calm morning; the lower part of the mountains and the lake were clear, but there was a thick fog above; the effect was very beautiful; the reflections were simply wonderful.

We stopped at the second narrows where the Eskimo camps were hoping to see something of the people. The camps and the kayak were still there, and it was evident that the Eskimos had been around since the time we passed; the things we had left had been taken, and we could see that the kayak had been moved since our last visit. It was very tantalising to find such traces of them but not to see the Eskimos themselves; however, we could not wait there on the chance of them showing up, so continued our voyage. We got two more caribou today, a cow and a large calf. Farther along the lake we saw another, a magnificent bull. He was on a small island and watched us come within a couple of hundred yards before he made any attempt to move. We did not try to get him; already we had as much meat as we could carry.

The weather went through some sudden and violent changes today from a lovely placid morning it turned to a stormy afternoon with strong northwest wind. The storm caught us at the western end of the lake there was no protection unless we crossed to the other side. Landing was out of the question on that exposed rocky shore; we had no choice but to paddle along through it as best we could, but fortunately it lasted only a couple of hours, then turned to a quiet sullen evening. We made camp that night at our old place at the western extremity of the lake. It had been a long hard day for us and we were both of us thoroughly tired, seldom were our sleeping bags more welcome or our tent more snug than tonight after the struggle against wind and waves we had just come through.

We slept till late next morning; then came the heavy work of portaging. We had cold weather on the divide, with frosts and snow and with the lakes

Sandy Creek

already beginning to freeze. The portage took two and a half days; we were camped twice on the divide. It was at one of these camps that we had the first real snow of the year, waking up to be surprised at the curious darkness of the tent and finding that this was due to a couple of inches of snow on it. At last we got back to Sandy Creek, and none too soon.

We made better time going down stream than we had coming up; there was more water in the river and our load was lighter. The Doctor usually took a pack and walked along the shore, while I took the canoe down. In spite of the lighter boat and the higher water there was still a lot of wading and hauling to be done, and wading was now getting to be a desperately cold job—not so much the wading itself as sitting wet in the canoe after.

The two-mile portage seemed child's play to us now compared to what

Simpson's Point

it had been on our journey up when we were fatigued and anxious. Our camp on both occasions was at the upper end of this portage; it was among some small spruce trees on a low sandy hill close to the river. We called it Simpson's Point, as there were traces of him here; and of all our pleasant camps in that country, I recall this as one of the most pleasant.

Although it was a mere rise it commanded a fine view in every direction. It was the only clump of spruce for some distance around; the country was undulating gravel plains covered with heaths and dwarf birch. It was a fine hunting–place; we saw many small bands of caribou on the surrounding plains, and we renewed our stock of fresh meat by killing a small bull.

After leaving the two–mile portage we saw lots of human footmarks on the sandy beaches of the river that had not been there on our journey up. We supposed at first that these had been made by Indians, knowing that the Bear Lakers had intended hunting caribou in this part of the country. At our old junction Camp all kinds of havoc had been wrought; trees had been chopped down and thrown across the river to make a kind of barricade. We could see from the character of the chopping that it must have been done by Eskimos, the trees had been worried through with some kind of a small and very blunt tool. I suppose that the barricade was a part of their deer–hunting devices.

We had left a cache here, some food in a waterproof canvas bag which I had hung on a cross–arm between two small trees. The bag was all right, but when I took it down I noticed that it was tied up differently from the way I had left it. When I opened the bag the first thing I pulled out was a small sealskin coat and a beautifully made pair of sealskin slippers! Besides these they had left a bunch of arrows with spruce shafts and copper-tipped bone heads and some trinkets carved out of ivory. We were puzzled to know what these things could be used for; it was not till next year that we found out. One of them was intended to shove through the nose of the marmots for convenience in carrying them; the other was a small handle with a loop of sinew attached for carrying the stomachs of caribou!

The only thing we missed from our camp was an empty lard pail; the things left by the Eskimos seemed liberal payment for that, but remembering a story of Amundsen's, I thought it possible that they might have discovered and appropriated the remains of the caribou that we had killed near this camp and the payment may have been intended to cover that as well.

Evidently they had been here until quite recently; hoping that we might yet see something of them, we camped a little above the junction of Sandy Creek with the main branch of the Dease River, intending to stay over a day;

I wanted also to make another trip to Granite Ridge to get some more bearings and connect up a compass traverse that I was making of our route.

We started out next morning; it was a hard, cold day, dull and cheerless with several degrees of frost; we thought ourselves lucky to be safely over the divide today.

There were signs of Eskimos all over the country now where there had been none at all on our journey up; in one place they had made quite an extensive camp on a hill, and the ground was covered with chips where they had been making sleigh runners.

We walked across that now familiar plain and climbed the Granite Ridge once more; it was a keen pleasure to look over the valley of the Dease River and recall how doubtful we had been when we saw it last, how nearly we had come to grief by taking a wrong road, and how successful our journey had finally been.

We saw some smoke far to the east of us near the river; we supposed it was Eskimos and even thought for awhile of making an attempt to walk there, but the distance was too great and evening was already closing in. We talked about going up stream next day to investigate, but that night we had more snow, and I don't think that even the certainty of meeting Eskimos would have induced us to wade up that shallow river again, hauling a canoe. It was a job we were heartily sick of, and in these low temperatures it was unpleasantly cold work. Moreover, we were anxious to get back to Hodgson's Point and see how Lion was getting on.

So we turned down stream next day; a bitterly cold day it was with a little snow on the ground, the small quiet bays of the river covered with ice and big collars of ice around every little twig trailing in the water.

We made good time; the river was high and we ran the same distance down stream today that had taken two days' hard work on the way up. But it was cold work; we still had to wade occasionally and it was a chilly job paddling in wet things in such weather.

Next day we made good progress and in the afternoon came to the long quiet stretch above Observation Hill; it was changed indeed since we were here last. Summer had ended suddenly and completely, the willows were now leafless, and the sedges withered to a dark yellow.

Such animal life as we saw was mostly on the move; the ducks had gone, the geese were going, every little while we would see and envy big flocks southward bound for another very different country that we too knew well. The ptarmigan, now nearly all white, were bunching up into big flocks and were very restless, making a kind of bluff that they would do the same as the geese if they wanted to. Only the muskrats carried on their business the same as ever, quite unconcerned by the general migration and the coming of winter; they were the true *habitants* of the country and the Dease River was good enough for them at all seasons.

We camped that night opposite Observation Hill and next morning climbed it again to connect up our traverse and look once more over the country that we were now more familiar with. We killed another caribou this morning close to our camp; like the first one we got it did not see us and came running over the plain right towards us; we waited till it got within easy range, then shot it. We camped below the canyon that night; there was quite a marked difference in the general advance of the season on the south side of this high ridge. At this camp we made a cache of nearly all the food we had left.

The nights were now getting quite dark and I saw the aurora tonight for the first time; it was extended right across the sky at right angles to the magnetic meridian. It was in the form of a spiral like a loosely twisted rope; there was no colour, it looked more like a slightly luminous cloud. A rapid movement was running through it from end to end, and I thought at first it was some kind of a cloud, some violent atmospheric disturbance, and for a few moments I was quite alarmed until I realised what it really was. The aurora became very familiar to us later, but I never saw it take this particular appearance again.

September 11th was the last day of our journey and it was over a part of the river that was the most difficult to navigate. Twice we broke holes in the *Polaris* today and twice we had to stop and repair her. One of these mishaps took place at the same rapid where we had knocked a hole in her when on our way up.

By noon we were at Indian Grave again; we stopped for lunch here, giving all our utensils an extra good cleaning and making everything on

*Hodgson's Point* and its Builder

board the *Polaris* as neat as we could; following the time–honoured fashion of the sea to make port with everything shipshape.

At last we came to the reach of the river along Hodgson's Point. We looked anxiously for the new house which we did not see when we expected to; what struck us most was the bareness of Hodgson's Point compared to our recollections of what it had been in summer. We ran the last rapid above the point and then our house came into view looking exceedingly neat and trim. Certainly Lion had made a fine job of it, and it was a contrast to the other ragged–looking shack.

We landed at a little wharf that Lion had fixed up right at the point; the *Aldebaran* was carefully stowed away ashore. No one was around; we

At the End of our Voyage

went up to the house, and then Lion saw us through the window; he was busy papering the inside with the pages of some magazines he had got from Hornby.

It was a joyful home–coming; we could take full satisfaction in the success of our trip now we knew that things had gone well with him also.

Lion was looking thinner and much bearded; he said both of us looked much sleeker and fatter than when we left and I have no doubt we did; certainly we had done our best to become so during the last ten days!

If the outside of our house looked neat and tidy the inside gave us a positive shock of pleasure. Lion had taken infinite pains with his work; the inside was snug, bright, and cheerful. I shall describe it later. His work had not been so exciting perhaps as our own, but just as important, and carried on, often under adverse circumstances, to the highly successful and creditable result we saw.

We had supper in the new house, Lion doing the honours. It was all very bewildering at first and especially strange to meet some one else and to hear another's voice. Then Lion's long beard had changed him so much in appearance that he looked to be scarcely the same person. The house seemed quite spacious after our little tent, and it was a consummate pleasure to take our meals again in such a civilised way, sitting on comfortable chairs at a real table and without any fear of upsetting things.

Altogether the evening was somewhat confusing; we had so much to tell each other that it was some time before we got a fair idea of our respective adventures.

After seeing us off Lion had gone ahead with his house building. On July 29th, the day after we left, the Bear Lake Indians arrived in their York boat and Father Rouvier with them. So we had beaten them handsomely in crossing the lake as well as coming up the Bear River. Hornby was following by canoe with some other Indians and the Father was awaiting his arrival to proceed to the Dismal Lakes on his evangelising mission. The Bear Lakers went hunting for caribou on the edge of the barrens, and their success so got on François's mind that he forgot all about his agreement, and Lion had considerable difficulty keeping him at all until some of the heavier part of the work was done. Hornby arrived on August 10th, and a couple of days after he and the Father made a start up the river. François left at the same time so Lion was now alone at his work.

About a week later Hornby and some of the Indians came back and the Father and the rest followed in a couple of days. Hornby and the Father had got as far as Observation Hill and they had returned to get some more supplies. The Indians intended to hunt along the southern shore of Dease

Bay; most of them were going to return to Fort Norman in their York boat, but the François and a few other families intended to remain in our vicinity all through the winter.

The Father and Hornby started up the river again on August 23d, and since that time Lion had been quite by himself, working at the house and hunting a little. He had got two caribou, and had dried some of the meat.

The season had been much warmer with him than with us; there is a very sudden change in climate when the high range of hills to the east of the lake is passed, a change that Simpson frequently remarks on and of which we had plenty of evidence later on.

## V

### THE TWILIGHT OF THE YEAR

I SPENT only a couple of days at Hodgson's Point, developing the photos we had taken on our trip. Then Lion and I started up the river again in the *Polaris* to make the most we could of the season before the river froze; we hoped we might find some caribou around Observation Hill and get a supply of meat for winter. I found this trip much more pleasant than the first one. The water was deeper, the canoe was lighter, and I was in a better physical condition. Best of all there was none of the incessant drive that had then been necessary. We could take things easy now and move or camp as we pleased.

We made the trip to Observation Hill in two days' easy work and put up our tent at the same place where the Doctor and I had camped on our way down the river. But though we hunted far and wide for several days we saw no signs of caribou.

Then we broke camp and paddled up to the end of the quiet stretch intending to investigate the East River, as we then called it, but it turned cold and we had heavy snow. The river might freeze up any night, so we decided to turn back to Hodgson's Point. We had lunch that day under the lee of a high sand bank; it was snowing heavily, and while at lunch Lion thought he heard the report of a rifle.

We made our way back to Hodgson's Point slowly, camping twice on the way and hunting as we went, but we saw no caribou, which was a great disappointment; the Doctor and I had seen so many on our journey from the divide that I had been quite confident of laying up some meat for winter. It was distinctly humiliating to come out on a hunting trip and be reduced to living on the remains of the caribou that the Doctor and I had killed a week before; the weather had been cold enough to keep the carcass fresh.

We had a final look around from the top of Observation Hill. Winter was now coming on in real earnest; it was a bright day but the snow on the hills was not thawing even in protected places in the sun; the plains were covered with snow and the small lakes frozen over. The distant mountains behind which lay the Dismal Lakes, mountains which the Doctor and I had looked at so earnestly on our first trip up this hill and which were now so familiar to us, were dazzling white.

We camped at Number One Camp on our leisurely journey home, and just before we got to Hodgson's Point we noticed some one walking along the shore, pack on back, and a couple of dogs loaded with packs following him.

It was Hornby; he had walked in from where he and the Father had been, and the rifle shot that Lion thought he had heard was Hornby's in fact. We got back to Hodgson's Point none too soon; the river froze over next day; we had utilised the season of navigation to the very last.

Hornby made his camp near our house and had supper with us. That night we had the first one of many pleasant evenings in our snug and comfortable little home, and Hornby told us of his and Father Rouvier's adventures.

It was the latter part of August before they made their way up the Dease River and though the water was higher by that time, they made very slow progress. Like ourselves they had only a hazy idea of the route to Dismal Lakes. They saw the traces of our camp on the sandy beach by Hanbury's Kopje and supposed without any question we had gone that way. They had a hard struggle to get to the lake, and by the time they got there the season was so far advanced they had to give up the idea of going on to the Dismal Lakes. Moreover, they met Eskimos in large numbers camped near the lake and hunting caribou in the country around, so they had nothing to gain by going any farther. We had noticed a small grove of spruce on the west side of the lake, the only trees for a long distance around. They decided to build a shack there and remain till the Eskimos returned to the sea. Hornby had come down to get his own shack ready for winter and to make arrangements with the Indians to catch fish for him. He had left the Father on quite friendly terms with the Eskimos and well fixed for food, as they had shot three caribou.

It was more than a month before we were settled down to the regular and systematic life that we afterwards led in winter quarters, but we got our house in better order at once and instituted regular duties. We divided these duties of the establishment into cooking, wood–chopping, and hunting, and took them in turn for a week at a time. On September 25th we started the new régime with myself as cook, Lion as wood–chopper, and the Doctor as hunter.

It was about this time that we saw most of the Bear Lake Indians; they came in from the edges of the woods where they had been hunting caribou and set up their teepees on the shores of the bay behind Big Island. They spent the next month fishing for white fish with nets and catching them in considerable numbers. These were the main food supply of their dogs.

When Lion and I returned from our trip up the river we found these Indians camped around Hodgson's Point, but to our relief they stayed there only a few days. Among them was the François outfit, quite unabashed and perfectly innocent that they had done anything out of the way in leaving Lion by himself. François had killed thirty–two caribou in the course of the season and brought us in a good supply of dry and pounded meat in return for the rifle we had lent him.

The total Indian population around our end of the take was then about twenty–five, men, women, and children; and this was the most we ever saw assembled in one place. There was François and his family, and François's brother Modeste and his family. Then there was an Indian that we used to call "Squiny," an evil–looking customer, though I must say his looks were the only thing we ever had against him. He had rather a nice, meek, and gentle wife, and several children. There was the Indian woman who worked for Hornby, and her family; I think she was a widow, I never knew her name. She was a cheery, energetic woman, quite as effective a worker as any of the men and a whole lot more cheerful. She had a boy of about fourteen,

Bear Lakers Breaking Camp

Harry we called him, who was a nice lad. Last, but not least, was an old chap we called Jacob, his wife and two boys.

Next to Mrs. François, whom we all liked for her quiet ladylike ways, Jacob was the chief favourite and most welcome at our house. He never abused our hospitality; he was always cheerful and nearly always amusing. He was the only one of the men who would condescend to smile and look happy.

These people used to come around our house a good deal at first, but found that we didn't want anything and that there was very little to be got out of us, so they soon went their own ways. The François family always considered they had some claim on us and any odd jobs we wanted done in the way of curing caribou skins, and so forth, was usually given to them, but some of our work went to Jacob too, who was really more reliable.

None of these people could talk English; such spoken intercourse as we

Some of the Bear Lake Indians. François Family to Right

had with them we carried on through Mrs. François, who could talk French. It usually took the Doctor's and my own combined knowledge of the language (little enough at that) to carry on a conversation. At first the Doctor used to tackle the job by himself, but Lion and I would laugh so much at his efforts, that he insisted on my taking part as well. It made a more efficient combination any way, for while the Doctor could understand much better what Mrs. François was trying to say I knew more French words than he did. Lion kept out of it and so had the laugh all by himself, and indeed it was laughable enough.

With the others we had to fall back on signs and the very few Indian words that we picked up. In the matter of signs Jacob was the most fluent. He would come in and sit down quietly in a corner, and after we had given him a cup of tea he would start making all kinds of curious and complicated flourishes. Usually we gathered from them what he wanted, but one day he had me completely at a loss to know what it was all about. He looked unusually mournful, his gestures were wilder and weirder than ever before; at last he gave up the job before my hopeless stupidity and sadly left the house. It was sometime later before I knew what he wanted on this occasion. He had had a row with his wife, and wished me to go and smooth things out!

I should say here that we saw none of that harsh treatment of their women supposed to be the custom of the Northern Indians. Hard labour by every one is the stern rule of the North, even the children and young dogs must help out as well as they can when the family is on the march. But we never saw that the women had any undue share of the work, or that they were treated as mere beasts of burden. On the contrary the sexes seemed to be on the basis of perfect equality in everything that concerned their domestic life.

What the Indians in turn thought of us I do not know, certainly we did things very differently from any white man they had seen before in that country, but what their real opinion was I have no idea.

The first two weeks in October were fairly mild, the season was milder with us than it had been with Simpson in 1838. The river, which had frozen across on our return, opened up enough to let us get down to the lake by canoe; so on October 2d, Lion and I made a trip to old Fort Confidence to have a took at those remains.

This place was built originally by Ritch of the Dease and Simpson expedition in 1837, and served as their quarters for two winters. It was reconstructed in 1848 and used by Sir John Richardson that winter when returning from his search for Sir John Franklin between the Mackenzie and Coppermine rivers; and again the succeeding year by Dr. Rae.

The buildings themselves were still standing in 1902 when Hanbury passed through Bear Lake but were destroyed by fire a few years later. Only their outline is now discernible and the four chimneys of the central building, standing like monoliths.

We spent a night there, pitching our tent right on the old site. It was

Camp at Old Fort Confidence

pleasant to get the extensive view of the lake again, the sun set wild and mournful that night, well befitting the season and the place. We returned to our own more cheerful quarters next day, calling on Hornby as we passed his house.

The weather soon turned cold, and by October 20th the river was frozen up, but very little snow had fallen as yet. As soon as the ice was strong enough to travel on Hornby intended to return to Lake Rouvier, as we now called the lake at which they had built their shack. Lion and I decided to go up with him; we wanted to get some experience of winter travelling and see how we would get on hauling our toboggans. These were the ordinary dogsleds of the country, short and heavy, quite unsuited for that kind of work but the only thing we had been able to get.

Hornby had two toboggans, three dogs to each and an Indian boy to help him. We hitched our own toboggans behind his, the ice on the river

The First Journey with Dogs

gave a splendid travelling surface, and though the lower rapids were open there was enough ice along the shore and snow on the ground to pass them without trouble. As we ascended the river we had constant evidence how much more severe the climate is to the northeast of Great Bear Lake.

The journey was a thoroughly pleasant one; we had fine weather, not too cold, the thermometer as yet going only to zero at night time. We made good time; it was a treat to pass at a swinging trot some of those places that had cost us so much time and labour when we made the ascent of the river by canoe. It took us only four days to make the trip in spite of having to follow all the windings of the river. The shallow head waters were frozen right to the bottom; Sandy Creek was solid ice clear through.

Above Hanbury's Kopje we came on the track of an Eskimo sled and pushed on fully expecting to find the Eskimos at the lake.

We found Father Rouvier looking well and cheerful, but learned to

The Shack Built by Father Rouvier and Homby

our disappointment that the Eskimos had all gone the day before heading north for the sea coast again. We spent a day with the Father, making a short trip over the hills to the north of the lake, but the weather was too thick to see much of anything. That part of the country was bad enough in summer; in early winter with the sun only a short distance above the horizon and the air full of frozen mist the outlook was miserable indeed.

The Father intended to return to Great Bear Lake at once with the dogs and the Indian boy. Hornby had got the Eskimos to make him a big sledge and the Father loaded this up with both the toboggans and a lot of stuff that he and Hornby had got from the Eskimos.

Hornby's plans were to remain at Lake Rouvier and try some trapping, and the Father intended to return a little later and bring him back.

Lion and I decided we would haul our own toboggans back to Hodgson's Point and hunt as we went.

October 31st, we broke camp and made an early start; the weather had been turning colder rapidly. Today was quite sharp, the thermometer had been –17° in the night. We had a pleasant breakfast with Hornby and the Father then started out across the lake.

We had about one hundred and twenty pounds on each toboggan. Our equipment was then by no means so well worked out as we had it by next spring. We were still using our Johnston's sleeping bags which we had lined with caribou skin; these were bulky and heavy and we had an unnecessarily large food supply with us. But we made good time on the ice, the tobogans slid along down stream almost by themselves. We camped that night below Hanbury's Kopje again; the Father with his big sledge and six dogs overtook us just as we got there.

He continued his journey down stream next day, while Lion and I went west to Simpson's Point hoping to see some caribou on those plains.

That place of pleasant recollections was very different now to what it had been when the Doctor and I camped there last. The hills and plains were covered with snow; there was a bitterly cold wind blowing and the air was full of rime. Of caribou there were no signs whatever, the whole aspect of things was so utterly bleak and miserable that we decided to make our way farther down the river. We moved our camp about five miles down stream that afternoon, and made an early start again next morning.

The days were very short now and it was necessary to camp about four in the afternoon. But one can always break camp in the dark, so we had to turn out very early to get in a fair day's work. We usually turned out at 4 A.M. or shortly after; an unhappy proceeding it was to turn out of our nice warm sleeping bags into a pitch darkness or at best unsympathising stars. It took about a couple of hours to get breakfast and break camp.

We made good headway till noon; then it began to snow. This freshly fallen snow made the hauling very heavy and our foothold on the ice more difficult than ever.

Things got worse and worse with us; we had to rest every few hundred yards, then every hundred, finally it would take a severe effort to move those toboggans fifty yards or less between rests. At last we could positively get no farther; we put up our tent on the top of a high bank of the river where there were a few spruce trees.

The wind rose strong from the northwest increasing to a regular gale the temperature dropped suddenly till it reached –4°. There was wood enough, but in that strong wind the fire did us no good; we got tea as well as we could and turned into our sleeping bags hoping the tent would hold, and for better times next day.

In spite of the gale the tent held all right and better times came; next day was perfectly lovely, bright, cold, and clear; we hauled those toboggans nearly twenty miles and camped that afternoon on the shore opposite the East River, near the same place where we had had lunch and heard Hornby's shot more than a month before.

We had intended to make another attempt to follow up this river and we did ascend it for a few miles, but evidently it ran through so much the same kind of country as the main branch of the Dease River that we gave it up as lacking in interest. It was on this little stream that Stefannson had made his winter quarters the preceding year; we saw no trace of them, but possibly they were farther up the river than we went.

There were no signs of caribou anywhere, and we were heartily sick of this toboggan hauling as a form of amusement so decided to get back to Hodgson's Point as quick as we could.

We got to the canyon that afternoon at 4 P.M.; this is about twelve miles by the river. The hauling had been fairly good and encouraged by a good tea we determined to try and make Hodgson's Point that same night. The moon was old enough to give a fair light, but it began to snow and the going became so heavy that by the time we got to No. I Camp, some four miles below the canyon, we were completely done up. We had not enough energy left even to light a fire or pitch the tent. We spread it out on the snow, unrolled and crawled into our sleeping bags, threw one flap of the tent over us, and fell asleep at once.

It was daylight before we woke up, a couple of inches of snow had fallen during the night, and we lay snug and warm under it, after one of the finest sleeps I ever had in my life.

The going was terribly heavy today; Lion had hurt his knee falling on a boulder in the river the night before and I had strained some sinew in my instep. It took us till one that afternoon to get to Indian Grave, and although we had only about three miles farther to go we had to abandon the heavier of the two toboggans here.

Even with both of us hitched to the smaller and lighter one we found it hard enough in our condition and we were thoroughly glad to get home. It looked unusually spacious and comfortable that night; our two candles seemed to give a perfect flood of light and a good meal at a civilised table was especially delightful.

The Doctor was well and cheerful; we heard that the Father had arrived only the day before. Notwithstanding the assistance of the dogs his trip had been almost as arduous as our own.

My week of cook had come around again, but it was a positive pleasure to turn out and light the fire after our last ten days' experiences.

The Doctor volunteered to bring in the abandoned toboggan; Lion and I agreed most cheerfully, and awaited developments. About 2 P.M. we heard him coming down the river, panting like a freight locomotive hitting a heavy grade, sweating like a horse, and smelling, as he said, like a sheep; he was wearing Stanfield's heavy wool underwear, and his description was a singularly apt one. After that experience the Doctor's opinion of toboggan hauling agreed perfectly with our own. I could express it truthfully only by offending that convention which permits more latitude in private speech than in a written record.

The Father made the journey to Lake Rouvier a few days after this with the dogs and toboggan, and the Doctor went with him. They met Hornby at Hanbury's Kopje. He had got so tired of the trapping job that he had made himself a little sledge and was on his way back to Bear Lake, when the others found him. I should add that after this Hornby's opinion of sleigh hauling was curiously like our own.

They all returned together; Father Rouvier and Hornby took up their quarters in Hornby's house and we settled into a regular and systematic life that lasted five months with scarcely a break.

# VI

## A WINTER IN THE ARCTIC

OUR house was about 14 ft. by 16 ft.; it appeared low from the outside but the floor was some six inches below ground level, and there was plenty of head room. The low floor was a good point under the circumstances as the soil was gravelly and there was no possibility of water draining in from anywhere.

It was built of spruce logs, "chinked" with moss and caribou hair, of which there had been great quantities lying around where the Hodgsons' outfit and their Indians had been curing skins.

It was thoroughly well mudded inside and out, and the inside was papered with the leaves of illustrated magazines that Hornby had given us. The floor was made of wooden blocks stamped in on end and all the cracks filled up with fine sand.

The roof was made of small spruce poles, chinked with caribou hair, then a layer of dry sand. Above all was spread the waterproof canvas we had brought for possible canoe–building purposes.

The fireplace was across one corner of the house; this most important feature was a regular triumph. Lion had departed from the usual custom of the country which makes the fireplace small and narrow, the logs to be burned standing on end; these fireplaces are an awful nuisance, besides being very inefficient. Lion's fireplace was a regular wide and deep one, with a big slab of quartzite for a mantelpiece. The chimney drew beautifully, smoke was quite unknown in our house.

151

We had two small windows that we had brought all the way from Fort Simpson; one of these looked towards the south, the other to the west. The door opened on the south side of the house at the corner diagonally opposite to the fireplace.

Lion had made a good table and we had four folding wooden chairs

Hodgson's Point at the Beginning of Winter

that we had got on the "Mackenzie River." Along the wall near the fireplace was a row of shelves; this was our kitchen. We kept the pots and pans on the lower shelves, the table utensils above them, and on top was a long shelf with a row of tin biscuit–boxes in which we kept various provisions: flour, sugar, rolled oats, beans, dried apples, etc. We had two cots; the original idea had been to make bunks, but instead of this Lion made a hammock for himself which he swung along the roof–beam. This was a good idea and gave us a maximum of effective room in the house.

Our meals were arranged on a schedule to give a minimum trouble to the cook, maximum daylight to the hunter and wood–chopper, and maximum comfort to all. We had breakfast at 9.30 A.M., dinner at 3.30 P.M., and a light supper at 8 P.M.

The Fireplace

For breakfast we had oatmeal porridge; occasionally bacon and beans, but more often some kind of hash made of caribou or ptarmigan with desiccated potatoes, bannock, and tea.

Dinner consisted of soup; caribou steaks or stews, or roast ptarmigan, with desiccated potatoes; bannock and stewed apples.

Supper was simply bannock and chocolate.

For our Sunday morning breakfasts we had coffee instead of tea and hominy instead of oatmeal. To the Doctor and myself the weekly coffee was a great treat, but Lion wasn't so enthusiastic about it; and I think that I was the only one who considered the hominy decidedly superior to the rolled oats. Sunday dinner was much the same as any other, but at supper that night we had maple syrup with our bannock.

The Kitchen

Our bannocks we baked in a reflector baker before the open fire and we actually got to like them better than the usual bread of civilisation as we found when we got back.

We all became quite expert as cooks, and the various soups, stews, and hashes we could make from caribou and ptarmigan with beans, desiccated potatoes and onions and erbswurst, would compare more than favourably with the productions of any highly paid chef. There was no reason indeed why they should not; we had the best of materials and what we lacked in mere experience was more than compensated for by our ability to approach the subject quite untrammelled by tradition; and I trust that our intelligence, at least, was of a higher order than that of the average professional cook.

At first there was considerable difference in the results of our various

Lion's Corner

cooking, but later on the Doctor's and my own approximated so closely in character that it was sometimes hard to distinguish our bannocks apart, or some of the various forms of hashes we made. Lion's cooking had an individuality all its own to the very last, especially his bannocks. I don't mean to say it was either better or worse than ours, but simply that it was different.

Lion certainly excelled in the matter of stews; the Doctor in hashes. I like to think that my own soups were distinguished for their peculiar excellency, but this is a daring supposition when those of both of the others were so exceedingly good. Our meals were always leisurely, and certainly we always did them justice. We all read at meals; our own stock of literature was very limited, but Hornby had quite a collection, cheap reprints of mostly good novels, which we read and re-read I don't know how many times.

Among the few books we had brought with us was Michelet's *History of France*, which I had borrowed from the Hudson Bay Co.'s factor at Fort Simpson. This served me for "breakfast reading" all the winter. I read that book through several times to my passing interest, but to very little permanent benefit. The Doctor also read it most assiduously. It was in two volumes and he would read one while I pored over the other; then we would exchange and re-exchange them. Whether he knows less about the early history of France now than I do I would hesitate to conjecture. I don't think Lion ever tackled this book; had it been in three volumes he might have done so.

At dinner time, in a more relaxed frame of mind, I always read some of the lighter literature we had. When we had gone through Hornby's books several times we tackled a heterogeneous collection of trash left by Hodgson: old magazines and various more or less lurid novels, dirty, torn, and with pages missing. Late in the winter when anything new was a real Godsend I found a quite simply and prettily told story called *Sunshine and Snows*; the front pages were missing, and to this day I don't know the author's name.

The hunter had no duties around the house, he was free to start out at once when breakfast was over. It was the woodchopper's job to dry the dishes when the cook washed them; as soon as this was done he too would leave the house on his day's work, giving the cook a clear field to clean up and attend to his various chores. These were usually finished about noon; by that time the cook would have water brought up from the river, wood brought in from outside, and all in the house clear for dinner. Then he was free to attend to his own personal jobs of washing, mending, and sewing.

The woodchopper would start out with a toboggan and axe, chop down the dry dead spruce that are always mixed in a greater or less proportion among the live trees in those northern woods. We each had our own particular districts which we were satisfied excelled in these dry spruce; we were always quite welcome to each other's districts, but generally stayed in our own. The first two days of the woodchopper's work were usually spent cutting down and hauling in the trees, and the next two or three in sawing them up and splitting them. It was always a point of honour for the woodchopper to leave a little more wood on hand than he found, so our wood–pile kept getting larger all the winter. Besides this each wood–chopper had his private stock of logs stacked up around the house ready to saw; a reserve against bad weather, or in case he should want to hunt. My own particular stock of logs was especially large; it never was used at all and it jars on me yet to think of some of those fine logs that I hauled in with such care and labour used by Father Rouvier and Hornby the following winter. About four or five days a week was all the woodchopper really needed to work; the rest of the time he usually hunted. There was a tacit understanding that the woodchopper should attend to any flocks of ptarmigan that came around the house but Lion was the only one who observed this faithfully.

The hunter usually started out about 11A.M., and made long trips in different directions, a different way each day. We had snowshoe trails strung all over the country for a radius of ten miles, and under our thorough patrolling we missed very little of what went on among the animals.

The commonest tracks we saw were those of wolves, wolverines, and foxes; sometimes we would get a glimpse of a wolf, but they were exceedingly shy. I saw a couple of white foxes together on one occasion; none of us ever saw a wolverine.

Caribou were really very scarce; only a few small bands left their tracks in our vicinity through all the winter months. Perhaps the Hodgson party hunting around here the previous winter may have frightened them away, or it may have been merely the chance of their movements, always very uncertain and capricious.

Bird life was more abundant than one would expect in that rigorous country, ptarmigan were plentiful all the time, usually in large flocks, and as a food supply came next to the caribou in importance. A few ravens were seen all through the winter; also hawks, apparently of several kinds, that preyed on the ptarmigan.

I saw a lovely big snow–white owl one day; it let me get quite close to it and I studied it for some time before it finally got nervous under my scrutiny and flapped away, showing a tremendous spread of wing.

Then we had whisky jacks, bold robbers they were. At first we liked to see them around, but they made such havoc among our supplies of meat, in spite of its frozen condition, that we were obliged to wage war on them. They were exceedingly pretty and fearless birds; I have known them come right into a tent where we have been at meals.

Last of all three chickadees kept us company right through the winter. We always plucked and cleaned the ptarmigan as soon as we got them, and kept them hung from a cross arm between two trees. Sometimes we had more than fifty ptarmigan on stock, seldom fewer than twenty or thirty; of course they froze as hard as rocks, but the chickadees pecked around them all the time and much preferred them to any of the other abundant scraps they had to choose from. We never missed what they took, but the voracious whisky jacks would soon pick a ptarmigan to the bones.

During the months of October and November ptarmigan was our only fresh meat, for those two months we killed on an average five ptarmigan a day. At the end of November we had a stock of more than fifty frozen ptarmigan all ready plucked and cleaned.

Lion was the most indefatigable ptarmigan hunter. We owed many a good supper of roast ptarmigan to him that neither the Doctor nor myself would have troubled to go after; he must have killed at least two–thirds of our supply. For hunting ptarmigan we used a Remington .22 rifle.

About the end of November a small herd of caribou came into our vicinity and we got six in two days. These were our main supply of fresh meat for the winter, and though we saw tracks of small roving bands they were few and far between, and we never shot any more caribou till the following March.

But we could always get ptarmigan, and always had all the fresh meat we needed. As Lion once said, "Our best friends in that country were the dead spruce trees and the ptarmigan."

Father Rouvier and Hornby got no caribou and seldom tried to hunt the ptarmigan. They had plenty of dried meat which they had got from the Indians and they caught lots of big lake trout through a hole in the ice. These were fine fish running from twenty to twenty-five pounds in weight and excellent eating.

We did not tackle the fishing proposition very seriously, much preferring to hunt caribou; but Lion made an attempt at fishing through the ice. His lines were set far in the bay, evidently not so good a place as near Big Island where the others had theirs, but that was too far to go, and he met with poor success. We had a stock of white fish that we had got from the Indians; these are fine fish, but none of us seemed to care for fish when we could get caribou hash, and if the cook sometimes served fish for a change the silence was usually so expressive that he finally got discouraged and our stock of white fish was used for dog food next spring.

By three in the afternoon the hunter was back and the wood-chopper had finished his work and cleaned up.

Dinner was the chief and most pleasant event of the day and the meal with which the cook took most trouble. As soon as it was ready he would light another candle; except from 3.30 P.M. till about 9 P.M., we used only one. Our eyes had got so accustomed to subdued light that these two candles always seemed a brilliant illumination.

The table was in a corner of the room alongside my cot; it was amply large, made of some old flooring we had got at Fort Simpson, part of the same house that our windows had belonged to. It had been painted first blue, then a coat of yellow on top, and enough of both colours was left to make the decorative scheme of our table cheerful and original. After we had finished dinner, the table cleared, and the dishes washed, we put on a tablecloth that Lion had made of some blue serge cloth he had got in trade from Hornby.

Then we would all write up our diaries and after that important job was finished, we played three-handed bridge.

We had only a couple of packs of cards with us, and they went through various stages of increasing dirtiness. After a few weeks' use, hearts would easily be confused with diamonds, or clubs with spades. Then it got difficult to tell any of the suits from each other, or face cards from spots; and the last stage was reached when it took close study to distinguish the backs of the cards from the fronts. Lion set to work one day and washed the cards carefully with a moist soapy cloth; the results were most gratifying, our cards looked as good as new to us and entered on another long spell of usefulness.

When Hornby was staying with us we played auction bridge, the Doctor and Hornby against Lion and me. Hornby was a brilliant, but somewhat erratic player, and to our occasional joy he never knew when to stop bidding. We even taught the Father to play "wisk" as he called it. He was very conservative in his play, and was so fond of finessing that it often led to astonishing results.

We also used to play chess after a fashion. Hornby and I would have great contests, both playing and working out chess problems. By his own account there was no problem that he failed to solve when at his own house, but on the journey up to ours he would somehow forget the moves. The Father played chess also, very seriously, and even worse than Hornby and myself. We had a small folding leather board at our house, the men were made of celluloid chips with the rank stamped on them. Hornby and the Father made a wonderful set for themselves out of wood; they alone knew the difference between a knight and a queen, and to our less experienced eyes it seemed that what were pawns one day were bishops or castles the next. This may have accounted for Hornby's success at the problems when he worked them out on his own board. Our chess playing may not have been scientific, but it gave us all lots of fun.

All November and most of December the weather was very uniform; the minimum temperature was about –20°. For a few hours in the course of the day it would rise some 10 or 15 degrees above the minimum. On

The Author in his Corner 161

November 11th, the thermometer went down to –36°, a minimum that was not reached again till December 22d when the temperature fell rapidly and the *grand froid* of winter began.

In November and December the weather was mostly dull and cloudy, with an incessant light snowfall which amounted to very little in the aggregate. At the end of December the weather turned clear and bright and was commonly so until the following July.

Our winter quarters were some twenty–five miles north of the Arctic Circle. We actually saw the sun for the last time on December 9th, and for the first time on January 1, 1912. It was not visible at our house between November 26, 1911, and January 9, 1912.

Christmas Day came on a Monday, and as Lion happened to be cook for that week he had the responsibility of the Christmas dinner. We had been reserving an Arctic hare for this occasion, the first and only Arctic hare we ever saw around the Dease River. Lion surpassed himself in making a plum pudding; he served this with blueberry jam that he had made in the summer. It was a most successful feast. Hornby and the Father were welcome guests, and after dinner we had Teshierpi toddy and a great game of "twenty–one," using squares of Peter's chocolate for stakes. We had served out an allowance of this all around, and Lion and the Father cleaned the rest of us out. In conclusion we had a grand supper of smoked caribou tongues, the great delicacy of the north and of which the Father was particularly fond. So closed what was really one of the most pleasant Christmas days that ever I spent.

After Christmas the weather turned very cold; the coldest spell we had was from January 9th to January 14th, when the minimum reading were as follows: –57°, –51°, –56°, –56°, –59°, –51°. From our experience I would call that climate in winter a rather placid one. We had high winds occasionally and drifting snow, usually at night, but I do not recall a single day on which the weather was bad enough to prevent the hunter making his rounds. The barrens to the east of us were evidently subject to much worse gales.

With the intense cold at Christmas the spruce trees were frozen solid and their stiff springy motion in a wind was very curious after the usual graceful swaying.

Our house was delightfully warm and snug; we let the fire out at night

Our Xmas Party. Left to Right: Lion, Father Rouvier, the Doctor, Myself.

time, but it was only in the coldest weather that the temperature would go below freezing point before lighting it again next morning. The ventilation was perfect; we controlled this with some holes bored in the door and wooden plugs; the air in that little shack was always as pure as the most fastidious hygienic crank could wish; indeed we were in this respect far better than in the most civilised houses.

Our water supply we got from the river; there was a deep stretch just off the point, between two rapids, and we kept a water–hole open in the ice all through the winter. At the beginning of April the ice on this stretch was over 6 ft. thick. We did not measure it on the lake; it may have been even thicker there.

On the whole the winter passed most pleasantly; I liked the regular systematic life we led and its varied duties. It was most agreeable, for

Hodgson's Point in the Middle of April

instance, after having turned out every morning for a week to light the fire and get breakfast, to lie lazily in bed and watch another man do it, and at last to get up leisurely when breakfast was all ready and the house pleasantly warm. Then to leave the table and let the others clean up the dishes, to put on one's snowshoes and start off for a fifteen–mile tramp.

One always began the hunting week with the highest hopes; week after week might pass without a sign of any caribou, but that did not prevent the hunter from having the most enthusiastic expectations the first day or so. Towards the end of the week one's confidence began to break down before

Lion

165

the persistent absence of any game; and at last, when one's woodchopping week began, it was almost a relief to start in on some real and productive work. There at least we got what we started out after; the physical exercise was varied and interesting, and called for a manual skill which it was a constant source of pleasure to exercise and improve.

After two weeks of exercise in the open one was quite ready to start in again on the more restful job of cook.

And so it went: each job was a foil to the other, each had its own advantages and came as a constantly pleasing contrast.

Since October we had seen very little of the Indians and from the end of November till the end of January we saw nothing at all. Then François and his brother Modeste came in one day and we learned that they had been hunting farther west. They had had a pretty hard time and some of them had lost their dogs by starvation. We gave them some supplies and they went back to where they had left their camp. We heard later that they got a fair number of caribou thereafter and were in no further want for the rest of the winter, though some of the others were not so lucky.

About this time Hornby and the Doctor decided they would try to make a trip to Fort Norman.

Two mails come into that post in the course of the winter, and one leaves for the south. The first mail packet arrives in January, and leaves early in February. The next arrives about the beginning of April. The mails are carried by dog sled and forwarded from post to post; letters only are brought in by these packets. We hoped to get some mail and to send out letters by the southbound mail packet.

Hornby had only two dogs; the prospects of their ever getting very far didn't look particularly promising to Lion and me, but they loaded up their toboggan and we saw them start from Hornby's house, watching them make their way slowly across the lake till they were out of sight behind Big Island. On January 27th, they turned up again, having found it too difficult a job with their limited dog power, and they had wisely turned back at the long traverse to Gros Cap.

A month later some of the Indians decided to make a trip to the post and Hornby went with them. They left on February 27th.

Our plans for the spring were now fairly well laid out; François had got

Father Rouvier

together a pretty good team of dogs, and he agreed to come and work for a month as soon as the Indians got back from Fort Norman.

Hornby was anxious to make a trip to the coast and we agreed to join forces. We hoped that he might be able to get some more dogs at Fort Norman, and we expected to have moved most of our stuff for the journey as far as the house on Lake Rouvier, by the time he returned.

About the middle of March, Lion and I decided to make a hunting trip up the Dease River again. A band of caribou had been in the vicinity and we had killed one near No. I Camp. We hoped to be able to lay up some depots of food for the dogs when we started the journeys to Lake Rouvier.

We had no dogs at all now, and had to haul our own toboggans again; but our equipment was much better selected; we took a lighter load and only one toboggan, so that it was comparatively easy work. We hunted around for several days in the vicinity of Observation Hill, but the only

Father Rouvier Leaving Hodgson's Point

caribou tracks we saw were nearer our own house. We had very cold weather on this trip, for two nights in succession the thermometer was down to −50° and with considerable wind at the same time. This was the only occasion when we had wind with such low temperatures; camping and hunting was a far from pleasant job under the circumstances. Our trip was not productive, but it gave us a certain amount of amusement and interest, and in spite of the severe conditions it removed the unpleasant impression left by our first experiment at sledge hauling and winter travel.

On March 24th, the Indians returned from Fort Norman; they brought a train of dogs from the Mission for Father Rouvier, who wanted to return to the post. One of the Indians was returning with him; Hornby had decided to remain at the post until they arrived and to come back again with this Indian.

The Indians brought us our mail; this was the red–letter day of the whole winter we now had news of our friends up to the beginning of November. How welcome those letters were, and how carefully read. I was lucky in having some correspondents who wrote regularly and gave detailed accounts of those doings that most interested me.

The Father left on March 28th. We were very sorry to say goodbye to him. He had added greatly to the pleasure of our life in winter quarters, and it was with sincere regret that we saw him off on his journey back to the mission.

# VII

## SPRING JOURNEYS

WITH the coming of milder weather and bright sunny days, our life in winter quarters began to seem rather lazy; it was time to wake up from our long winter's sleep and to start on our travels again.

François came on April 4th and pitched his teepee close to our house; we had arranged that his brother Modeste also should make one trip to Lake Rouvier.

On April 6th, the Doctor, François, and Modeste started on the first trip; they had two toboggans, with three dogs to each, and a total load of nearly 500 lbs. on the two toboggans.

I went as far as the canyon with them; travelling was most pleasant now, the surface of the snow was at its very best, the days were long, bright, and placid, and the temperatures pleasant, neither thawing nor too cold. They took only six days to make the round trip, but with the good condition for travelling they were able to make a shorter route over the plains, avoiding the big bend of the river. They were all suffering from snow–blindness when they got back; they had glasses but didn't use them until too late. This was the only occasion that any of our party were troubled in that way.

Modeste couldn't make another trip for us, he had his family to look after: François wouldn't, although we were looking after his family for him so far as we could make out he objected to going into a country where there weren't any trees. After considering the matter for a day or two he agreed to lend us his dogs, and on April 14th the Doctor and Lion started up the river with another load.

On April 17th, Hornby arrived from Fort Norman; he had been able to get only two more dogs, which was rather a disappointment, but we hoped

Lion and the Doctor Start for Lake Rouvier

to get enough from the Indians to make up two teams. He also brought some mail that had come by the second winter packet, and we got news from the outside world up to the beginning of January.

The Doctor and Lion got back on April 19th, after a pleasant and successful trip. The sun was getting so strong now that they were both much sunburnt.

We had still a considerable quantity of stuff to send up and decided that Hornby and the Doctor should make another trip; the Doctor to remain at the house on Lake Rouvier until we joined him. On April 22d they started;

we had five dogs now, and on this trip the toboggan was loaded with over 500 lbs. of stuff, the heaviest load we ever hauled on one toboggan.

Hornby took five days to make the round trip; he actually came back from the lake in two days only, but he was sitting on the toboggan all the time and his five dogs took him along at a fast trot. He got back on April 28th.

François' Tepee at Hodgson's Point

April 30th was the day set for the grand start.

The François family were still around Hodgson's Point and on the day before we started some of the other Indians came in too. They had been hunting on the north shore of Bear Lake and had suffered pretty hard times, losing many of their dogs by starvation, and one woman had died.

Hornby had decided to take the boy, Harry, along with him; this gave him a travelling companion and assistant, and left him free to follow his own devices when we got to the Coppermine in case he wanted to go on to the sea at once, or to return to the Dease River before the snow melted.

We now had seven dogs altogether; the little I have said about dogs hitherto must not be taken as a measure of their importance in the general

From left to right: *"Potash," "Punch," "Geoff," "Nigger"*

scheme of our life; I have simply been waiting an opportunity to do them the best justice I can.

One's experience of the dogs of civilisation is quite inadequate to give any idea of how a dog's character may develop and how acute his intelligence may become under the stern conditions of life in the North. Hard work, strict discipline, and rigorous treatment develop character in dogs no less than in men. Certainly I had never before seen dogs of such pronounced individualities as those we had now got together or who in their traits and behaviours so exactly resembled certain types of men.

"*Potash*" was the senior dog of the crowd. He was big, rather thin and spare, but wiry and strong. He was a wily and knowing old brute; he

The Unwilling "*Potash*"

understood what was said to him to an extraordinary degree and knew the meaning of every move we made about camp. His memory for places where caribou has been killed was most lively; he loved to go back and visit the remains no matter how much he might have eaten in camp; in consequence, we had to keep him tied all the time we were camped. He never did any more work in harness than he had to, but he was a valuable dogs as leader; he would obey the voice and go as directed, consequently we usually put him as the leader of the first sleigh. It was always very funny to see Potash coming to be harnessed up. At first he would pretend not to hear when he was called, and mighty well the old dog knew when he was called to be harnessed up and when any food was being handed out. When the sudden deafness was of no further avail he would creep along stiffly and painfully, his head bent down, and his tail between his legs, one foot dragging slowly after the other, his whole attitude humble and suppliant, every movement betokening extreme reluctance and bodily weakness.

"*Punch*" was the next in seniority; by seniority I don't mean age, but general importance. He was a big handsome yellow dog, always cheerful and friendly, always willing and working his hardest. In camp his behavior was perfect; he could be absolutely trusted not to steal anything, no matter how tempting and accessible it might be.

Potash and Punch were Hornby's two dogs; they had always been great friends with us at our house, indeed they liked it much better than their own. Potash particularly was always most unwilling to leave the place.

"*Cuchar*" was one of the dogs we had got from François. He was a great big black dog with long hair, and eyes like a bear; he was strong as a pony, the strongest of all the dogs, and the most valuable as a worker. He was rather inclined to be savage at first and we had to be careful not to be bitten when harnessing him up; as time went on he got tamer, but he never showed any particular affection for us. He was as stoical and as undemonstrative as his late master. He was a most inveterate and bare–faced thief; we had to keep him tied up all the time, but he would make the most daring use of such few opportunities as came his way. Many a beating he got for his thievishness, but he didn't mind beatings in the least. I have seen one of us take a stick to him for having stolen something and Cuchar would be trying to steal something else while he was being punished.

"*Geoff*" belonged to Mr. Leon Gaudet at Fort Norman. He was a most handsome dog, black and dark yellow, with long hair; in character he was much like Punch whose half–brother he was, ever cheerful and hard working, a most lovable, playful dog, always up to some mischief of a more or less innocent kind. He was a hard worker, rather fond of fighting, and while he wasn't so strictly honest as Punch he could be trusted to behave himself around camp if we let him loose.

"*Nigger*" was a smaller dog, black with a few white markings; Hornby had got him at Fort Norman; he had worked for Hornby and Melville before. He reminded us of a certain type of men–little runts, little mentally, usually physically as well–men who are always irreproachable in their conduct, always working hard, always immaculately good; whose whole life is an example. Just such a virtuous and irritating little prig was Nigger. Punch and Geoff were both good dogs, but they were thoroughly lovable characters and never made a parade of their goodness as did that insufferable small–minded snob Nigger.

"*Husky*" was another of François's dogs; he had got him in trade from the Eskimos in the summer. He was the typical "husky" dog, grey, with sharp upstanding ears; he was a curious character and had little in common with the other dogs. At first, with his pretences of being a great fighter, he had them all, except Cuchar, bluffed into paying him a certain amount of deference. His favourite trick was to come charging down on them when they were eating and send them spinning, then appropriate their fish, or whatever it might be. From the way he took it up at once with our dogs it was no doubt a trick he had acquired and worked successfully amongst the Indians though I never saw him try it on Cuchar. He tried it with Geoff one day, and as soon as Geoff got over his astonishment, he sailed into Husky and would have killed him if we hadn't interfered. That was the last time Husky tried his charging game with any of our dogs. He was a fairly willing worker when hauling a toboggan, but his legs were short in proportion to his body and he couldn't travel as fast as the other dogs. He was almost useless when we came to "pack" the dogs later; his short legs made it very difficult to load him and he was a most unwilling worker when it came to carrying anything on his back. He was a bad thief, though not in Cuchar's splendidly daring style, and we had to keep him tied all the time.

I have said he was always a kind of outsider with the other dogs. If we were in the tent and any of the dogs were up to some mischief, Husky would give the show away by growling loudly; the chastisement that the offender usually got was always to Husky's evident great delight, and he was always on the alert against anything doing that he could not take part in himself.

It would be an injustice to Potash to put any other dog in the first place; as I can't give *"Jack"* that honour I have left him to the last. If Nigger reminded us of the self-righteous Pharisee, Jack was like one of these stout, easygoing, imperturbable men, faithful and reliable, the kind that never make a fuss, whose work is usually well done, and in a way that gives themselves least trouble, and who don't work overtime unless they can't help it. Jack had been with Hornby and Melville the preceding year and had been given by them to the Indians. We made Jack's acquaintance when the Indians were camped at Hodgson's Point in the autumn; he

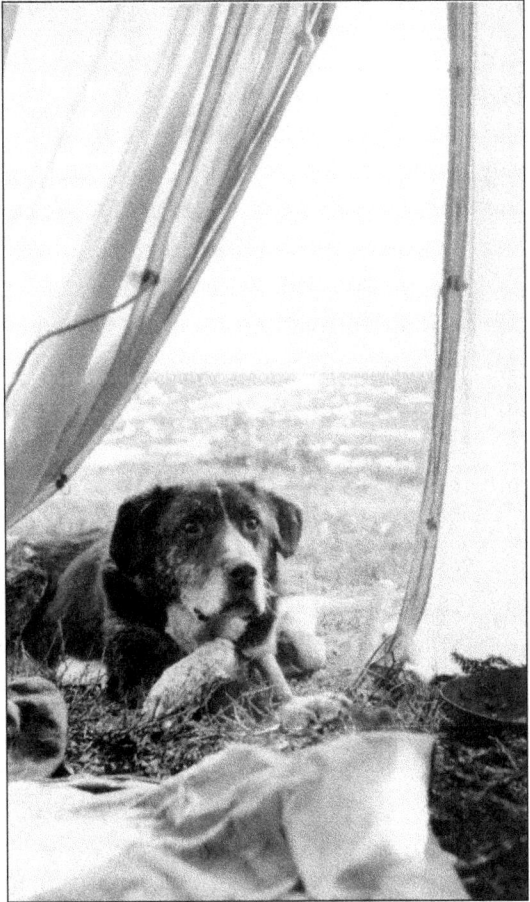

*"Jack"*

had always been a prime favourite with Lion and me to whom he particularly attached himself. He spent the winter with the Indians, and we never saw him from the end of November until just before we started on the trip, when he came to our camp with the Indians on the last day. Poor old Jack was in such a condition that we hardly recognised him; he was so starved that he could scarcely walk. We rescued him and took him along with us. He could only follow behind the toboggans for the first few days, but picked up wonderfully quickly, and in the course of a few weeks of good feeding became one of our best dogs. He attached himself to Lion and me more than ever. Certainly he was always my special favourite, I don't know why. He wasn't so lovable a dog as Geoff or Punch, such a good worker as Cuchar, or so clever as Potash; in mere point of intelligence he was one of the most stupid of the dogs, as he was one of the ugliest in appearance.

But so it goes, and so unreasoning are we in our affections.

Harry joined our party the night before we left, and the Indians all went off again to their camp on the lake. Harry was a rather nice lad and though he was of very little real use he was usually amusing and certainly added to the general liveliness of the party.

At first it was quite evident that he didn't think much of white men and their ways. It is true he knew Hornby well, but Hornby had been so long in the country that he had got to be almost like an Indian himself in his way of living. Harry got to know us better later on, and I think ended in even admiring us!

We had everything loaded up on the toboggans ready for an early start; all our things were carefully put away in our little shack, which would now have to take care of itself.

The last breakfast was quite sad; it was the last civilised meal that we were likely to have for some time and we were sorry to leave the place that had been such a comfortable home to us. Harry was particularly solemn through it all; we had hominy that morning as a special treat, something that he had never seen before in his life, and he tried to eat it with a fork plainly contemptuous of white men and their ways, especially their food.

Everything was made as secure as possible in the house; we closed up the door, trusting to find all well on our return.

# VIII

## TOBOGGAN AND PACK

ON April 30th, at 9 A.M., the start was made; we had two toboggans with three dogs to each, Hornby took one toboggan with Punch, Geoff, and Cuchar, I took the other with Potash, Nigger, and Husky. Harry ran on ahead; Lion was the hunter and poor old Jack trailed along in the rear.

The weather was fine and bright, our toboggans were not too heavily loaded and the trail good; travelling was a fair delight under such circumstances and after the confinement and the small worries of preparations that had taken all my time for nearly a month. We camped that night above East River, and the next on the overland traverse that we used to make between the Dease River and Lake Rouvier.

In passing over the barren plains next day, we sighted a large herd of caribou, the largest herd we ever saw, numbering perhaps several hundred. Luckily we saw them ourselves before the dogs got wind of them. Lion and I went ahead while the others looked after the dogs. The caribou were scattered over an open plain; they were in a very difficult place to approach and we had to open fire at an extreme range. Fortunately we got three of them; as soon as we had settled the caribou, Hornby came up with the toboggans, Harry and the dogs in a great state of excitement. Cutting up caribou was Harry's particular delight; he always used to appropriate for himself the bones of the fore legs and crack them for the marrow.

The dogs ate till they could positively eat no more. Jack had such a

179

feed as he had been stranger to for a long time, and his rapid recovery dated from these first caribou we killed.

We loaded up most of the meat on the toboggans; it made them very heavy and between this point and Lake Rouvier we had the heaviest going,

The Grand Start

as the wind had swept the snow off the exposed ridges; in some places we had to double up the teams and haul the toboggans over one at a time.

We got to Lake Rouvier that same afternoon and found the Doctor all well. He had killed several caribou also while we were away, and had brought in most of the meat on his back, so we had plenty of fresh meat now; the dogs fairy revelled in it; we gave them all they could possibly eat, and treated ourselves the same way.

The country between Hodgson's Point and Lake Rouvier had become so familiar to us by this time that it seemed only now that we were

making a start. It required four toboggan loads to take all the stuff we
wanted at the Coppermine River, so that we would have to make two
journeys.

We spent a day at Lake Rouvier giving the dogs a rest and a chance
to fill up. Lion and I also had to dig out Hornby's canoe; it was buried in a

Crossing Wind–Swept Ridges

hard packed snow–drift at the edge of the lake. We intended to use it on our
return journey down the river and wanted to make sure that it would not get
damaged when the ice broke up.

It turned mild very suddenly that afternoon; we had a strong south wind
and the temperature rose to 40°. This was quite unexpected and very unwelcome; we hoped it might turn cold again at night, but it kept mild and the
warm south wind blew all through the night.

Hornby and Harry had been sleeping in a tent; the Doctor, Lion, and
I in the house. The roof leaked so badly that Lion and I took refuge on

the comparatively dry chips scattered outside; the Doctor was luckier in the location of his bed.

We started at seven next morning; the snow was soft and the going very heavy; the ascent of the long grade at the northern end of the lake was a hard struggle. The summit of this divide between the Dease River and the Dismal Lakes is a narrow level valley extending for several miles; along it lay a few small lakes. On the east side of this valley the hills rose precipitously; there is a conspicuous high bluff that we had hitherto called "Hornby Hill" but which we identified afterwards as one that the Doctor and I had seen from the top of Teshierpi Mountain the summer before and had then called "Dorothy and Marian;" from that point it appears as two somewhat similar peaks.

The ice on the small lakes was already covered with slush, so rapidly was the snow melting under the warm strong south wind. To the north of the divide there is a lake large enough to be conspicuous; the Doctor and I had seen this also from Teshierpi Mountain, and had then called it "Mountain Lake," but it was not till later on that we identified it.

We travelled over the surface of this lake for several miles, then shaped our course by compass for where we expected to find Teshierpi Mountain. Leaving the lake we crossed a high ridge; from the summit of this we had a grand view of the great depression in which the Dismal Lakes lie. The first landmarks I could identify were the long line of precipitous cliffs running between the Coppermine River and the Dismal Lakes; neither the Doctor nor myself could recognise Teshierpi Mountain for some time, and small blame to us. For we had seen it only from points northwest to northeast of it from which sides it appears round and regular in shape. Its southern aspect is quite different, being a series of precipitous basalt terraces like the Coppermine Mountains, of which it is in fact the most southerly spur.

We followed the undulating ridges down to the Teshierpi River; the hauling was very heavy, the snow was disappearing with astonishing quickness in that brilliant sunshine and warm wind; it was a rarely beautiful day and the view ahead of us was singularly lovely under the bright and tranquil sky.

We crossed a few more small lakes; their shape on this side of the divide was generally long and narrow; their surface was six inches deep in slush.

Teshierpi Camp

Our toboggans swished along through it like boats and I blessed the good equipment of waterproof bags that we had.

We got to the Teshierpi River and followed that down to a point on the southern slope of the Teshierpi Mountain about four miles from where the river discharges into the narrows between the second and third Dismal Lakes.

We made camp on a small rise above the river; such snow as the winds had left on that exposed spot was completely melted off. It was the first dry ground camp we had this year, and we found it an indescribable luxury after the nuisance of camping on snow.

Teshierpi Mountain rose high to the north of us; there is some small stunted spruce on this southern slope so we had plenty of firewood. Moreover we had the unwanted luxury of plenty of good drinking water instead of having to melt snow over the fire, which was always troublesome and which always tasted nasty.

We thought Teshierpi Camp the acme of comfort; even when we camped here again more than six weeks later, and after we had become decidedly fastidious in the matter of camp sites by the uniformly good fortune that had been our lot in this respect, we still thought it a pleasant place. There is in fact something curiously attractive about all that country at the eastern end of the Dismal Lakes, an attraction that we had felt the summer before and which Hanbury seems to have experienced no less than ourselves.

The rapid disappearance of the snow was very awkward for us, and we decided to bring the second loads of stuff to this point at once before the sleighing got any worse. Fortunately it turned colder again next day; the Doctor and Hornby made the trip back to Lake Rouvier while Lion and I prospected ahead for a good route. We wanted if possible to strike right across to the Coppermine Mountains instead of following the Kendall River. We also climbed Teshierpi Mountain again and I was able to check up the bearings I had taken in the autumn and to identify the landmarks to the south.

Next day the others returned early, the freeze–up had made the travelling good and the second trip had been a comparatively easy one.

We decided to do our travelling by night henceforth; the conditions were more likely to be better and there was practically daylight all the time; within a week from this time we were unable to see any stars at midnight.

The next two days were very stormy with high winds and a little fine snow; it was not till the evening of May 8th, two days after Hornby's and the Doctor's return, that the weather moderated enough to allow us to proceed on our journey.

On May 8th, Lion, Hornby, Harry, and I made the first trip with two toboggan loads of stuff, leaving the Doctor to begin his geological work around Teshierpi Mountain. We started at 9 P.M.; it was a raw blustering evening, but turned finer as the night went on. A short distance below our camp the dogs scented a caribou and got so excited that we had to upset the toboggans to hold them back, while Lion went ahead and shot the caribou. It was a little yearling bull, the first bull caribou we had got this year. We took most of the meat along with us; we had now become so expert in butchering caribou that we lost only half an hour over the whole business.

We crossed the second Dismal Lake, then made a straight line for the western bend of the Coppermine River. The ground was smooth and level, and though there was only a scanty covering of snow, the sleighing was nevertheless very good; we had those ideal conditions of a hard crust with a little loose snow on top. We crossed an undulating plain with small lakes here and there; we kept parallel to the ridge of the mountains on the northwest and our route lay over several of these lakes.

In some places there were small scattered spruce trees, mostly dead; we stopped for lunch about 2 A.M. on a little oasis of dry ground with plenty of dry spruce for a big fire. In high spirits we sat before that cheerful blaze and filled up on caribou; enthusiastic over sledge journeys, over night traveling, over the country, and most of all over caribou liver.

A few miles farther on we reached a high point from where we could see the most westerly sweep of the Coppermine River.

The banks are very steep along this bend, but we found a place a few miles above Stony Creek, where we got down all right with our heavily loaded toboggans. I was surprised to find how the snow had gone from the familiar southward–facing hills; the terraces of the mountains were mostly bare except where huge drifts lay at the bottom of the basalt cliffs.

The river had been completely flooded, though now it was frozen up again. The surface was smooth ice that would have been fine for skating.

Just below the place where the Doctor and I had camped the summer before we came on an extraordinary sight.

The country through which Stony Creek runs, just before it joins the Coppermine, is a curious elevated delta, forming a level triangular space among the mountains about half a square mile in area. A considerable part of this plain is made up of a regular bed of boulders, very uniform in size

The Rock Slide in Spring

and in one place this peculiar " boulder bed" extends right down to the shore of the river, a cataract of rock a hundred yards wide that we usually spoke of as the "Rock Slide." Even in the middle of August there was an accumulation of ice on the boulder bed; it looked like a miniature glacier and we were puzzled then how to account for it.

Now the Rock Slide was like a vast frozen waterfall, a veritable Niagra of ice. Hell Gate was filled almost to the top of its walls with a steep frozen river, and the boulder bed beyond was covered many feet thick with bluish-white ice. Evidently Stony Creek must run all the year round from some subterranean source, and these huge accumulations of ice are the results of its constant overflowing and freezing as it overflows.

As we passed it we looked at the old place where the Doctor and I had been located the previous summer, but it was not a good situation for our

First Spring Camp at the Coppermine Mountains

camp now. With some difficulty we made our way up the steep ice incline of Hell Gate and pitched our tents on a bare grassy space at the top of the bank between the boulder bed and the river. "Boulder Bed Camp" was the name we gave the place; it was to be our base camp and we set up our tents as carefully and comfortably as we could. It was a good place in that there was plenty of smooth level ground around the tents and a fairly abundant supply of wood from a grove of stunted, scattered spruce close to the camp. It commanded a fine view of the river, but it was terribly exposed as we found to our sorrow more than once later.

By the time we had the camp fixed up it was noon. Lion and Hornby turned in for a sleep, while I made a trip to the very top of the Coppermine Mountains, hoping to have the same success hunting now as on the first day of the former visit; but I got nothing except an Arctic hare. The summit is a rocky table–land; its thousand feet of elevation had made a consider-

The Coppermine River in Early May

able difference in the progress of the season; the thaw had taken very little effect on those drifted plains, a dreary desert of rock and snow.

Hornby started back to Teshierpi Camp that same night about twelve; Lion and I spent the next day hunting and building a platform ready to cache our stuff when we started farther on our journey. The others returned the following morning; the travelling had been good and we found ourselves happily located at the very place we wanted to be and at the earliest time at which we could profitably begin investigations.

Indeed we had got over the latter part of our journey none too soon;

only the next day after the Doctor and Hornby arrived with the last load it turned so mild that all snow on the levels disappeared like magic. Travelling by toboggan was over for the season.

The small creeks were running high and the water flowing deep over the ice along the shores of the river. We had not expected the spring to start so early as this; for while we hoped that a spell of frost might even yet set the ice on the river hard enough for us to travel on. For a week we did not lose hope of this, but at the end of that time it was quite evident that all possibility of travelling on the ice was over.

In fact the weather continued so warm and the ice in the river seemed so likely to break up before long that we began to think seriously of making some kind of a craft, though we knew that on both of Simpson's journeys he had been unable to descend the river with his boats until nearly six weeks later than this.

There was enough dry spruce to build a small raft, but a raft in that swift river among ice would have been too unmanageable, in fact I should not care to attempt rafting down the lower Coppermine under any circumstances.

Our toboggans were of no more use to us, so I broke them up, splitting them into thin strips and from these strips we made a canoe frame, using the curved part for the ribs and lashing them all together with *babiche* and electric insulating tape of which I had brought a roll. We intended to cover this frame with a silk tarpaulin.

Considering the materials at our disposal I was quite proud of the *Good Hope*, as we called her. It made a great impression on Harry; he said that he supposed white men couldn't make anything unless they had a hammer and nails, and he started to make a funny little model to take back and show the other Indians how we had built a canoe of our sleighs. No sooner had we started building the canoe than spring came to a standstill, the weather turned raw and cold with severe gales from the northwest. We felt the exposed condition of our camp now, and had to build brush windbreaks to protect our tents. That which Hornby and Harry rigged up in front of theirs excelled in the variety of its materials and the picturesquesness of its appearance, a result attained, not by sudden flight, but the sum of successive inspirations; there was no underlying central idea consistently worked out.

It may have been a lucky thing for us that the condition of the river

The "*Good Hope*"

remained so impossible for any kind of navigation that we never even launched the *Good Hope*. Perhaps the Eskimos have found and used her since. I can imagine their amusement and contempt over the results of which I was so proud, but which would compare very unfavourably with their neat workmanship, although they have no better materials and far inferior tools.

In the meantime the Doctor was carrying on his geological investigations and we were all prospecting and hunting. Lion had shot a small bull caribou the first day after we camped.

The skins of the caribou were now lined inside with the larvae of warble flies or bots. At this time these were great white maggots as big as the first of a man's little finger. They grew no larger, but as the season progressed they changed in colour to a dark yellowish brown.

These warbles eat their way through the skin and drop out on the ground

Hornby's Triumph

where they go through their final transformation to a fly something like a big bee.

Besides the warbles we sometimes noticed at this season a mass of worms in the back part of the caribou's throat: two wriggling bundles as big as eggs of loathsome things like pale yellow caterpillars. Harry always showed a great dread of them, and said that the dogs would die if they ate any of these worms.

What a tormented life those wretched caribou must lead! harried by wolves, hunted by men, tortured by flies, their throats sometimes full of writhing worms and their hides punctured by warbles with a hundred holes.

Lion's caribou lasted us only a couple of days and we were beginning to need meat badly again when I was fortunate enough to get four fairly large bulls. Hornby, Harry, and I had started out one evening and about

Warbles in the Skin of a Caribou. Early May

three miles from camp we came on a caribou in a wide muskeg. We were a long way from it and I saw only one at first. Then another appeared. As I shot at them two more came into view and I got them all. Harry was in a state of the wildest delight; Hornby had recently given him a big knife and here was a fine chance to use it.

We took as much meat as we could carry back to the camp, intending to get the dogs and pack in the rest of it. It was midnight by this time and the Doctor and Lion had turned in. When we got back with our loads of fresh meat the dogs made such a racket in their excitement that it woke them up. Lion volunteered to help us, but the Doctor had already been working hard all day and didn't see any fun in packing caribou meat to camp all night, so he turned in to sleep again. We had a good supper of fresh heart and liver before we started, and while sitting in our tent eating it Hornby thought he heard Husky giving one of his premonitory growls and yelled

Hornby and Punch

but to Geoff to behave himself; with his playful disposition Geoff was usually the offender in these cases. The growling continued; at last Hornby went out to see what was going on and found that the noise came from the Doctor snoring in the other tent!

We spent the rest of the night bringing in the meat; this was my first experience of packing with dogs, and the amount of weight they were able to carry fairly astonished me. Cuchar could carry as much as I would like to attempt.

A few days after this the Doctor and Hornby made a trip with the dogs about ten miles farther down the river and located another camp there, "Camp Basalt," from which the Doctor continued his work for a few days

By May 24th we had given up all hopes of the ice going out of the river soon enough for us to use the *Good Hope*, so we decided to make the journey on foot, packing the dogs.

We first took one load of stuff to a point twelve miles below Camp Basalt; "Camp Melville" we called this, after Hornby's former companion. We left the Doctor here to geologise while we returned to our base camp at Boulder Bed, then joined the Doctor at Melville again with another load.

Then we moved slowly down the river prospecting as we went.

Packing with dogs was necessarily a slow and laborious business, so much time was needed to straighten out the packs and to arrange the loads on the dogs that it usually took us four hours from the time we got up in the morning until we were ready to start. Cuchar, Punch, and Geoff were the best dogs; they all carried loads of more than fifty pounds. Jack would only carry about forty; if we put any more on him he would lag behind. But Jack was a very reliable packer, especially when crossing streams, and we could put some of the things on his back that required more or less careful handling. Potash was not so strong, but willing and careful. He always carried the tents and a tarpaulin; these altogether weighed thirty pounds. Nigger was too small to be much good as a packer and he was generally a nuisance, always slipping his pack and most skilful in assuming a look of complete innocence when it was done, and in making it appear that he was not to blame but that the fault lay with the person who had put it on his back. Husky was the worst of all; he was short in the legs and his back sagged like the cables of a suspension bridge. At best he could carry only twenty–five or thirty pounds, and he was so much slower than the other dogs that he always lagged behind.

Each of us had our own packs of clothes, blankets, etc., including rifles, ammunition, instruments; our total loads averaged about 70 lbs. each. Hornby and the Doctor were the best packers; Hornby was a small man but very strong and wiry, and the load he could carry was remarkable. I have known him carry a load of 125 lbs. of caribou meat a short distance into camp. I think that I was probably the poorest packer of the lot.

The weights given above may not seem very great when one recalls the familiar way some popular writers speak of 150–lb. packs as though they were mere trifles. I should emphasise the fact that we were strictly average men and that all the figures I have given were obtained by actually weighing and not merely guessing or estimating.

If I were to *guess* the weight of a 70–lb. pack after having carried it ten miles I should say "about 300 lbs.!"

The order of our march was usually as follows: I led the way with the faithful Jack close at my heels, later on Potash showed a decided preference to walk behind me also. Then came Hornby, the Doctor, and Harry, behind them the rest of the dogs. Lion brought up the rear, the hardest and most thankless job of all; Husky always lagged behind and had to be incessantly urged along with a stick, howling, yelping, and whining; a ceaseless futile protest that we were all going much too fast. Husky was an awful nuisance on the road; later on we never loaded him at all and even then he wouldn't keep up.

The time we spent between Boulder Bed and Melville was the most unpleasant part of our packing experiences; it was the roughest country we had to cross, the ground was very wet, and we still had sharp frosts occasionally. The dogs' packing harnesses would get wet and then freeze: it was a troublesome job thawing them out in front of the fire before we started. Moreover we did not have the details of packing, the arrangement of packs, etc., so well arranged as we got them later on.

At Camp Melville we had one day of such bad weather, with high winds and fine driving snow, that we had to stay in camp.

Then came a change, sudden, complete, and delightful. The weather turned bright and warm, beautiful beyond description, and though big drifts still lay here and there in the hollows the country soon dried up and the walking was generally very good. Our next camp below Melville was a particularly pleasant one, the spruce was getting very small and scant now, and occurred only in small clumps along the river. We called this "Camp Brulé"; it was near a small grove of spruce, part of which had been killed by fire at some remote time. Hornby objected to the name and wanted to know why we could not use English; considering the number of Brulés one

Burnt Camp

finds in the North, his objection was good, so we changed the name to Burnt Camp. And the name came near being prophetic; we very nearly did lose our tents by fire spreading through the moss.

We spent two days here. Lion got four fairly large bull caribou close to the camp, and I got three at a place ten miles farther down the river, so we had a supply of meat ready at our next camp.

We made our way down the river by easy stages. The spruce trees got smaller and scantier, finally we left them behind altogether, and had to use

Sandstone Cliffs on the Lower Coppermine River

heaths, willow twigs, and mosses for cooking, but everything had dried up so quickly in that constant bright sunlight that we never had any difficulty in making a fire.

The country through which we passed was a succession of undulating

The "*Musk Ox Rapids*"

plains over a sandstone formation and between well–defined parallel ranges of basalt hills. These lay nearly east and west at right angles to the course of the river, which cuts through them, usually forming rapids at the intersection. Such rapids are the "Sandstone" and "Escape" of Sir John Franklin, and most notable of all the "Bloody Falls" of Samuel Hearne.

The ranges decreased in height as we went farther north; judging by the eye alone one would never suspect how great is the difference of elevation. The Coppermine Mountains rise 1100 feet above the river, the next well-defined range about 500–600 feet; while the last range, through which the river cuts at the Bloody Falls, is less than 200 feet high.

We first saw the sea from the summit of the second range; it was about twenty–five miles distant in a straight line.

Characteristic Sandstone Formation, Lower Coppermine River

We had fixed on the Bloody Falls as the location of our farthest north camp; on our last day's march we kept expecting to see them long before we had any real reason to do so, and at every summit or rising ground we hoped to get a nearer view of the sea. Our noon camp that day found us on the brow of a sandy height; we had come nearly nine miles already and we expected surely to have a sight of the sea from this summit, but instead of this we overlooked a wide valley bounded on the north by yet another range of hills. As a matter of fact we were then just above the

Escape Rapids. We could not tell from here where the river passed this range, but it was evident that the Bloody Falls were still eight or nine miles ahead of us.

A striking feature of this range was what at a distance we took to be

A Basalt Dyke. Sir John Franklin's *Sandstone Rapids*

some vast snow drifts almost level with the top of the ridge; as we got closer we found these were in fact a great mound of white clay and sand with some smaller detached ones.

Even when we got to the range we saw nothing of the Bloody Falls, the river curved past the big sand mound making a cut bank that was quite impassable; our only road was to cross the range and we were all too to tired attempt it that night. So we put up our tents on its southern slope in a corner between the basalt ridge and the smallest clay mound. This clay mound looked exactly like an immense tailing dump from some ore–concentrating mill and we called it "Tailing Dump Camp."

There are several of these peculiar clay mounds on the eastern side of the river as well; they are a curious and striking feature, and I am surprised

Tailing Dump Camp

that so little has been said about them by former explorers. Only Sir John Franklin mentions them and that very shortly, comparing them to icebergs, which is a very good simile. Richardson speaks of the rocks at Bloody Falls being "covered to a depth of six or seven hundred feet" with a bed of "greyish white rather tenacious clay." The "six or seven hundred feet" is an extravagant over-estimate and illustrates what I have just said about the apparent height of these successive ranges and how curiously deceptive they are to the eye.

But travelling as we did along the heights above the river gave us a much better chance to observe the country than making the journey by boat, and these white hills no doubt looked far more striking to us than they had to our predecessors.

The Bloody Falls

After supper we climbed the ridge; like all these basalt hills its southern aspect is sharp and precipitous while it stretches away gradually to the north. From the summit we got a wonderfully beautiful view of the sea under a gorgeous sky, the high western capes and the many islands distorted to curious and ever–changing shapes by a remarkable mirage. We could also see the Bloody Falls, a mile or so farther down the river.

It was 10.30 P.M., when we returned to our camp, the sun was still above the horizon, fiery red, the mirage had become stranger and more fantastic than ever; of the real appearance of the islands in the gulf and the land to the west we could form no idea whatever.

Early next morning we started over the ridge again, heading for the Bloody Falls; it was a fine bright day, and the view to the north was now

The Bloody Falls—a Closer View

undistorted by mirage; for the first time we could see what the islands and the coast to the northwest really looked like.

As we were looking over the river from a high point we were delighted to see some Eskimos on the other side, so made our way down to the Bloody Falls as fast as we could. These are not really falls, but a swift crooked rapid; the river cuts through several hundred yards of basalt running between perpendicular walls of sheer rock in some places not more than fifty yards apart. There is a drop of about fifteen feet in three hundred yards, the river rages violently through that narrow crooked channel; it is a regular sluice, impetuous and turbulent. The rapids continue a short distance below the gorge; in fact the greatest drop and the swiftest water is just beyond the basalt walls.

The Lower Part of the Bloody Falls

The river then flows more quietly between high banks of sand and clay.

The Eskimos saw us before we got to the rapids, and hailed us with as much pleasure as we did them; there was a great shouting of "Teyma!" and waving of arms.

At the particular moment of our arrival the upper part of the rapid was clear and the lower part jammed with huge blocks of ice. The river had started to break up in earnest, more and more ice coming down all the time.

The Eskimos were on the opposite side to ourselves; they were soon joined by a number of others that came over the hill from some camp below

until there must have been twenty–five or thirty of them, mostly women and
children.

After a while two of the men ventured across to our side, walking over
the jammed ice; it was a most risky proceeding on their part, as the whole

Drawn by W.J. Wilson from a Photograph

Eskimos Crossing on the jammed Ice at Bloody Falls

thing was liable to be swept away any moment. These were some of the men
that Hornby had met the preceding summer; they were quite friendly and
apparently delighted to see us.

Presently one of them went back across the ice, and returned to our side
again carrying the skin of a musk–ox; a third man came with him, the Doctor
thought that this was the same man we had met at the end of Dismal Lake.
They were all nice looking men; one was a particularly fine looking fellow,
several inches taller than the others, active, robust, with rosy checks and an
air of alert intelligence.

It was a delight to meet these vivacious, well–bred people after the sulky Indians; their manners indeed were just as good and very similar to our own. We could carry on a conversation only by signs; they were very clever at this, even old Jacob's exploits were thrown into the shade by their brilliantly expressive gesticulations.

They explained that they had come to spear salmon, giving a most comical imitation of a fish wriggling at the end of a spear. As soon as the flies came they were going inland to hunt caribou; and the representation they gave of a man pestered by mosquitoes, slapping his face and neck, was extremely realistic; one could almost hear the mosquitoes buzzing. They did it all with inimitable gravity, quite engrossed in their efforts to make us understand.

We wanted also to find out where they picked up the pieces of native copper used by them for their weapons and utensils; they pointed to the south and gave a ludicrous representation of a man bending under a heavy load to typify a long journey, finally struggling up a steep hill and arriving quite exhausted. Certainly their fertility of resource and invention at signs was extraordinary.

Meantime the rest of the Eskimos watched us from the high rocky bank opposite. One of the women, evidently the wife of the biggest of the three men and really quite attractive in appearance, was much concerned when her man crossed the ice, and well she might be. When he got back safely she welcomed him with many signs of affection. They had one little mit with some bead work on it; probably it had come to them from the Indians, and apparently it was some kind of an amulet or charm; they would exchange it every once in a while; he left it with her when he crossed the ice jam.

They went back to get some more musk–ox skins, of which we understood they had seven altogether at their camp; they had scarcely crossed before the whole ice jam carried away and we had no further intercourse with them.

The Doctor wanted to make some geological notes, Hornby hoped that the Eskimos might be able to get across the river again, Lion and I wanted

to go on to the sea, so we left the others and struck across to the north–
northwest, in which direction we expected to find the shortest route to the
shore.

We walked over a gently undulating grassy plain, with occasional small

Drawn by W.J. Wilson from a Photograph

View Towards Cape Hearne

lakes and muskegs; this part of the country was very wet, and in some
places we had to wade long distances up to our knees. The distance to the
sea was greater than we had expected; we were tired and hungry, but after
having come so far we did not like to forego the honour and glory of getting to the actual shore itself. At last after a weary nine–mile walk from the
Bloody Falls we stood on the very edge of the continent.

The beach is sand and mud; there is a considerable quantity of driftwood along the shore; I was surprised to find so much, but it is all small
stuff. I saw nothing but spruce.

The ice was fast to the beach; there was no kind of a tide crack, nothing at all to indicate any rise or fall of tide, though the driftwood was considerably above the level of the ice.

On Coronation Gulf

Probably the wind has most to do with the level of the water in this part of Coronation Gulf. There may be considerable changes in this level during the short summer season when the ice is broken up, or when there is open water in the gulf itself, in the surrounding straits, and in the ocean to the west. But when the surface of the sea is solid with ice there is evidently very little change in level.

In all the small lakes the ice had thawed leaving a space of open water around the shores, and I do not know why this was not the case with the sea ice as well. The latter must set later in the autumn, and the reasons for the ice thawing first around the shores of the lakes apply equally well to the sea ice also, but we could walk from the gradually sloping sandy beach right on to the sea ice.

It stretched smooth and solid to the north, the surface was mostly hard level, there was a little water and slush on it in places but only in very shallow pools; it would have been fine to travel on just then.

We unfurled the flags we had reserved for this occasion and took pictures of each other proudly standing on the ice of the Arctic Ocean, the northern limit of our long journey.

It was "Homeward Bound!" now, our faces were turned south again at last; a retreat ever southward, of which the end was five long months and many thousand miles away.

We returned by much the same way that we had come; about three miles from the sea we saw an Eskimo on the side of a small hill hunting ptarmigan with bow and arrows. We waved and shouted to him, but he never paid the slightest attention to us. Then we saw there was a small Eskimo camp not far away with some others standing around; I suppose the man thought we were waving to them.

We crossed a swamp to this camp; it was a small lean–to of caribou skins stretched over some poles, the people were apparently on the march and this was merely a temporary shelter.

They welcomed us warmly; there were five in the camp when we got there: an old man and woman, evidently the father and mother, a girl of twelve or thirteen, a boy about ten, and a little girl of seven or eight.

Presently another girl came into camp with a bundle on her back, and later on the young man whom we had seen hunting ptarmigan.

These were decidedly the most pleasant of the Eskimos we saw; indeed it was hard to believe, so far as conduct and behaviour went, that we were not dealing with highly civilised and cultivated people. They had the same easy manners and the same well–bred ways usual with all people of culture.

At first we tried to carry on the conversation with the old man, but his

Eskimos Near the Coast

wife shoved him aside in a good–natured way, as much as to say, "Leave this to me, I am far cleverer than you." The old lady was very voluble and kept up a steady stream of talk, as though she was quite confident that we understood perfectly all she was saying. The eldest girl took part in the "talk" in a quiet way; she was a nice girl, really quite charming, though it may seem a curious application of the word, and the next eldest was a pleasantly cheerful little thing.

As far as we could gather from them, they had come from the North

(Victoria Land?), that they had been catching seal, and that they were going to hunt caribou. They knew that the other Eskimos were at Bloody Falls. We did not know of Stefannson's whereabouts and tried to find out from these people, but could not make out whether they had seen anything of

A Summer Camp

him or not. They had a large sheet–metal trough and a couple of tin pots, perhaps these had been brought in by Bernard in the " Teddy Bear."

The young man kept much in the background; the women seemed to be of most consequence in this family.

I had written down a number of Eskimo words in my notebook, a kind of a little dictionary that I had made in the winter from a French–Eskimo dictionary by Père Émile Petitôt. They understood very few of these words, no doubt because of my own imperfection of pronunciation, but once in a while I would get out a word that they did know and then their astonishment and delight was most amusing; they would crowd around the book and listen as though they expected to *hear* something from it. I wrote down a few words that I got from them and this seemed to surprise them no less; they all wanted to try their hands with the pencil; it was delightful to see their joy at being able to make marks that to them probably looked much the same as my own. The eldest girl was particularly interested in the book and in trying to write.

They wanted us to have something to eat; set over a very small fire was the sheet–metal trough I have mentioned. It was full of some kind of a stew, but though we were both very hungry neither of us felt inclined to tackle that slimy repulsive mess of luke–warm oil, blood, and half–raw meat. Our food had always been of a fairly civilised order, and one needs a little breaking in to stand this sort of thing, so we declined their hospitality as gracefully, I hope, as it was offered.

It was late when we got back to our own camp; we crossed the ridge almost at midnight, the sun was still above the horizon when we were on the summit, and once more marvellous mirages transformed the rocky coast line to vague enchanted shapes; a fairy land of gold, crimson, and mother–of–pearl.

Both of us were tired and hungry; it had been an exciting day, and we had walked about twenty–two miles since a very meagre lunch. We speculated whether any of the Eskimos had found their way across the river; it was really a relief to us that we found none of them around our camp when we arrived at last. Hornby woke up as we were having supper and joined us; they had seen nothing more of the Eskimos and the river was now an impassable barrier.

We started southward again next morning, it was almost a matter of regret to do so; I should have liked nothing better than to have spent the summer with the Eskimos and become familiar with them, but this was quite out of the question under our circumstances.

We moved south in the same leisurely manner that we had come north, making the same day's marches and stopping at the same camping places.

The weather kept delightfully fine, the sun never set, the country was familiar to us, and we could pick out the best route; it was drier and better for walking; best of all our loads were not so heavy and steadily got lighter day by day.

We spent a couple of days at Burnt Camp again; this had been a kind

Camp Comfort

of outpost camp where we had left a cache of food and some of our things; there was still a lot of caribou meat left and we were all well fed. We abandoned a lot of our stuff here, all our fur clothing, except a capôte that Hornby kept, and our fur sleeping bags, of which we were obliged to take a sorrowful farewell. Another treasured article left to its fate at this camp was an aluminum reflector baker that had done good service, but we had no flour nearer than Hodgson's Point now. These things and our furs had never gone farther than Burnt Camp anyhow.

From this camp we struck right across the mountains, avoiding the big bend that the river makes; we made a new camp not far from our old Camp Basalt. We called it Camp Comfort and never was a name better deserved. It was on a little level grassy point, a small creek ran close to our tents, and we were surrounded by spruce

Boulder Bed Camp in June

that seemed quite big to us after the scrubby trees we had got used to.

We got back to Boulder Bed on June 12th; it was quite like getting home to return to this familiar place. We even had one of the severe northwest gales that seemed to blow with peculiar violence in this corner of the mountains, and our windbreak assumed proportions larger and more elaborate than ever.

A few days more were spent here; the Doctor had some final geological notes to make; the rest of us were hunting and making ready for the journey back to Lake Rouvier. We got two more caribou close to the camp, which were very welcome, especially to the dogs. They had a good rest and all they could eat; they were in splendid condition now. As a matter of fact we were all of us at our very best.

A lot more stuff had to be abandoned here: one of the tents, a small sheet–metal tent stove that had been our best friend in the early days of the trip; clothes, tarpaulins, ammunition, snowshoes, waterproof canvas bags, and a few tools; these things had all served their purpose. We had even a considerable quantity of food left, mostly corn meal, erbswurst, and salt pork; our caribou hunting had been successful beyond expectation.

We cached all the stuff carefully as Hornby had quite made up his mind to spend the summer and another winter in this country, and he intended to come back in the autumn as soon as the sleighing was good and haul this stuff to Lake Rouvier again.

Saturday, June 15th, was the day of our departure, and a dull gloomy morning. We said goodbye to a camp where our experience had been on the whole so pleasant and successful, and struck across the mountains in a straight line for the narrows between the second and third Dismal Lakes. A sandy bar extends across the lake there; we intended to ford it at that point.

We reached the summit of the long ridge between the Coppermine River and the Dismal Lakes, "The Palisades" we had called it. From here we had our last view of the Coppermine River and the Stony Creek valley.

On our return from the Bloody Falls we had found the swift lower reaches of the river free of ice, though the shores were lined with stranded floes, and with masses of crushed ice shoved far up on the banks by jams. Along this part of it the current is not so swift, and the river was still covered with ice; indeed there was very little change in its appearance now from what it had been three weeks before. The ice was very rotten though and likely to go to pieces completely at any hour. Thus our experience of when this river is likely to be sufficiently free of ice to permit navigation checks very closely with Simpson's.

Today's march was one of the longest we had made. When we came in sight of Dismal Lake it was evident that the water was so much higher that our plan of fording the narrows was quite impracticable. The ice in the lakes was still fairly intact, though there were wide spaces of open water around the shores. We hoped to be able to cross on the ice somehow, so

Glacier Cove

shaped our course for Glacier Cove where the Doctor and I had camped the preceding autumn, and where we knew we would find some spruce.

We got to that familiar place in the evening; one of our tents we put up at the identical spot on which it had been pitched nine months before; we even drove our tent–pegs into the same holes, so little change takes place in this frozen country.

But we didn't think Glacier Cove so pleasant a spot now as when we had camped there before. There was none of the lovely colouring that had made it so delightful on the former occasion, and the view of the ice–covered lake was not a cheerful one. The season was not nearly so far advanced here as it had been on the Coppermine River.

The ice looked so rotten and so likely to break up if a high wind should come that we decided to cross as soon as possible. The Doctor wanted to

Ferrying our Stuff to the Ice

look over that country to the northwest which we had been foiled so often in examining before. Hornby, Lion, and I hunted around for dry spruce logs out of which to build a raft. Fortunately we found a few trees big enough. Our knowledge of the country gained last year certainly came in useful to-day. We had another good illustration of how one's eye may be deceived by custom: the logs we got looked quite big to us and the raft seemed amply large to carry one of us; when we came to launch it we realised how small these trees really are; the *Dolphin* as we called it (one cannot flatter a raft by the use of the feminine) would scarcely float myself. I made a precarious voyage out to the ice taking a heavy fishing line with me that Lion luckily had; with this line we pulled the raft to and fro between the ice and the shore and soon had most of our stuff ferried over ready for a start when the Doctor came back.

It was after seven in the evening before he returned; he had made a

Crossing the Second Dismal Lake

long trip to the northwest and had seen recent Eskimo footmarks. No doubt it is by some easier route to the west of the Coppermine River that these people find their way to the Dismal Lakes.

In the course of the day the ice had moved so close to the shore that it was possible to get on it at a point a short distance from the camp; we ferried the last loads over on our raft and got the dogs across at the point, except Jack, who wouldn't follow the others, but insisted on swimming out to me.

We hauled the raft out of the water and loaded all our stuff on it; we had built it something like a sleigh, with the intention of dragging it across the lake as we might require it on the other side. It took the united efforts of the entire party, men and dogs, to start that load; the ice was terribly difficult to walk on, the surface was melted by the sun into long sharp needles set at an angle like the teeth of a saw. The *Dolphin* slid along pretty well when we got it started, for the slant of the ice needles was in our favour.

Teshierpi Camp Again

At a long hook–shaped point on the other side, the ice was so close to the shore that it was not necessary to use the raft; we broke it up and made a bridge of it from one ice–floe to another and then to the shore. The *Good Hope* had been both sleigh and canoe, but the *Dolphin* went one better and in addition to these was a bridge as well.

We loaded up the dogs and headed over the northwest shoulder of Teshierpi Mountain for our old Teshierpi Camp. It was a heavy task climbing those steep grades; we had all been working hard during the day, the Doctor more than any of us. As we crossed the shoulder of the mountain

I got a bearing of the sun at midnight to check up our time. There was only one watch in the party now that would run at all.

A heavy shower came up from the south; we could see the clouds shutting out the hills with a blanket of rain, and when we got at last to our

Eskimos at Teshierpi Camp

old camp site it was raining heavily; with the usual perversity of the weather this lasted only till we got our tents up.

We took it easy next day. It was five in the morning before we had been able to turn in and we slept till afternoon. Lion went off hunting and met two Eskimo men that he brought into camp. These were not up to the standard of the other Eskimos we had seen in either looks or intelligence. We had no fresh meat so gave them some bacon; they liked this raw well enough, but didn't seem to care for it cooked. They drank tea with apparent satisfaction, sugar was evidently quite distasteful to them, yet the first man we had met the preceding summer had certainly enjoyed milk chocolate.

Harry was very officious as cook and particularly anxious to impress on these men how important a person he was in our party. We gave them a

Eskimo Hunting Ptarmigan

few odds and ends, a tin bowl, and a nickle–plated metal fork apiece. They gave us to understand that their camp was not far away, so I went back with them. They were very slow walkers; I was walking certainly not more than three miles an hour, but this pace quite distressed them; they were panting heavily and the sweat rolling in streams down their wide glistening faces.

Their camp was in fact three or four miles from ours; it was situated on a gravel hill overlooking the second Dismal Lake, near the first narrows.

There was an oldish woman and a younger one, and a small boy; they were all ugly and not nearly so clever and bright as the others we had seen.

They had a small wedge tent about 6'x6' made of caribou skins with the hair on. They had a kayak frame with them, the cover had been taken off, no doubt so that they could carry it more easily in a wind.

It was about 10 P.M. when I was at their camp; the sun seemed so bright to me that I was quite mis-led when taking some pictures and under–exposed them badly.

We turned out early next day; the Eskimos all came on the scene as we were having breakfast. I took some more pictures of them; they stood up and posed quite as I required them. As soon as I had got what I wanted they all turned around to see some-

Looking Pleasant

thing that interested them in the camp, so I got another almost equally inter-esting view!

They had brought a lot of stuff that they wanted to trade: low sealskin shoes in stacks, lines made from the thick hide of large seals, and a kind of parchment made of young sealskin. Evidently all this stuff had been got ready in the winter, especially for trading purposes with the Indians. The sealskin shoes were far inferior to those worn by themselves; we had

Harry and the Eskimos

seen some of the lines used by the Indians, who value them highly for sleigh lashings. What the parchment was for I do not know.

They were much dismayed when they found we didn't want anything, but we had all we could carry already. We gave the younger woman a fork like those we had given to the men, and to the old lady a large spoon. But she didn't want a spoon, it was a fork that she had set her heart on and was so anxious to have one that she offered stacks of sealskin shoes for it, then a big white wolf skin, and was finally ready to give almost anything she owned. But we hadn't any more forks to give away. I found out the reason why she wanted a fork so badly, they prized them as *combs*; the kind of comb they make is narrow with long teeth and does in fact look rather like the lower part of a fork.

Hornby's fur capôte attracted her attention next; it wasn't a good one their own fur clothes were really much better. She put it on and went

Kayak Frame and Double–Bladed Paddle

through what I suppose was meant for some kind of a dance; she pranced around like a heavy draught mare in a sportive mood. The exhibition was truly absurd, so ridiculously unfit for dancing was the old girl. But no less preposterous are some of the exhibitions one sees nowadays in a country that is proud of its enlightenment, nor was this grossly silly old Eskimo dame one whit more extravagant in her folly than some "civilised" women I have since seen, whose sense of the fitness of things has been completely obsessed in their infatuation over a fashionable craze.

We said goodbye to the Eskimos and on June 18th we started on the last and longest of our day's journeys with packs.

It was hot and fatiguing ascending the long grade to the divide, and for the first time this year we had mosquitoes in large numbers, though they

Our Visitors

were very feeble as yet. We kept to the east of the route followed when we travelled with the toboggans and found a good way over the hills to the left of Mountain Lake. At last we reached the downward slope on the other side of the divide and once more we got a distant view of Lake Rouvier far below us.

As we got near the lake the difference in the progress of vegetation was very marked; on the south side of these hills the season was several weeks farther advanced than it had been around the Coppermine River.

The dwarf birch were coming out in leaf; what a welcome sight were those first green and tender leaves of the year! The mosses were all shades of greens and browns; again we were reminded of some brilliant carpet by the close–packed masses of colour we walked over, but now the general tone was green.

And a Back View of Them

The lake was still covered with ice, though it was very rotten and there was a wide water space around the shores. At the edge of the spruce woods on the western slope we saw some Eskimo tents and a number of musk–ox skins spread out over the big boulders, but there was no one in sight.

Everything was all right at the little shack; there were signs that the Eskimos had made a large camp close to it since we had been here, but nothing had been disturbed. Even an axe, an article of inestimable value to them, was just where it had been left.

It was late by the time we had tea; directly after Lion and I went to examine and launch the canoe. She was all right too, though the seams were sadly opened out and she leaked like a very sieve. We filled her with water, rather she did that herself and only too quickly, and we left her in a little bay to soak.

It was arranged that Lion and I should take the canoe and most of the stuff down the river, while the others walked with the dogs. About four in the afternoon of the next day we made a start; nothing had been seen of any Eskimos up to the time we left.

It was a perfect joy to get into a canoe again, clumsy and leaky though she was. We thought of our last experiences coming down this river when hauling our toboggans, and contrasted this delightful mode of travelling with the constant worries and fatigues of packing with dogs. We coasted along the shore of the lake and entered the river; it was much lower than I expected to find it at this time of the year, and it was necessary to wade. We fell back into our old way of doing things very quickly: Lion carried a heavy pack along the shore and I hauled the boat over the shallow bars. But this was sport compared to hauling a toboggan, and better times would surely come. The camp we made that night, without the nuisance of the dogs, seemed positively ideal to us.

The first two days of our journey were raw and cold with a high north wind. It was necessary to wade in many places on the upper reaches of the river and this was a miserably cold job. Heavy rains one evening made things worse. I hadn't even a coat, my fur capôte had been abandoned at Camp Melville. Lion had always used a Mackinaw coat throughout the journey, and if I had the laugh on him a few times in the early part of the trip, it was his turn now. But these things didn't worry us; it might have been worse: we might have been hauling a toboggan, or staggering along under a pack. We sat in our snug tent over a steaming stew of wild goose that Lion had killed that afternoon, shot with a rifle on the wing, and thought of the camps we had made in snow and darkness, contrasting with them our present happy state.

One after another we passed the familiar landmarks: here was Notman

Dyke, here was the high bank where we had spent a particularly miserable night the November before. We had to stop every now and then to patch up our canoe; if we struck a rock, no matter how gently, it knocked a hole in her, she was so frail and battered up after the hard usage of the trip that Hornby and Father Rouvier had made the previous summer.

Our last camp on this journey was just above East River; we were both so stiff, chilled through from sitting in wet clothes all one cold windy day, that we could hardly crawl ashore.

Then times changed for the better; the weather turned warm and placid once more, and navigation became easier. A great flow of water was coming down the East River, and the main Dease River was so high below the junction that no more wading was necessary.

Our last day on the river was beautiful; bright, warm, and calm; a delightful contrast to the turbulent weather of the last two days; the quiet stretch of the river above the cañon was never more peaceful and lovely than to-day.

I thought of our various former trips over this, the fairest part of all the river; how we had seen it first in its very prettiest summer dress and at the very height of its summer activities. A couple of months after and it was late autumn, the banks frozen and ice forming along the shores, with the leaves all gone, the grass brown, the sedges withered and mournful, the birds hastening south, and over all the stern menace of the swiftly approaching Arctic winter.

I thought of the time we had come down this same stretch, painfully hauling our toboggans. Winter had now come in stern reality with darkness, desolation, and death. And again, four months later, when we hunted over these stark solitudes in vain; the thermometer at nearly –50°, the river buried in vast drifts of snow, swept by the winter gales from the barrens above; the reign of death long established, final and complete—Thou fool! that which thou sowest is not quickened except it die; life is indestructible as energy or matter, a manifestation indeed of both. Here was life triumphant; exuberant and rejoicing; the river had resumed its ever-changing sky–dyed surface and its tranquil flow; the grass and the willows were greener than ever, and the dwarf birches more delicately beautiful. The muskrats were alert and jubilant; geese flew overhead, ducks circled at every bend, and ptarmigan ashore chirped to their just hatched little ones; on every side there was activity and exultation.

View on the Dease River

Below the cañon the river now ran a wild and frantic career, but we were too pleased at nearing home to let any rapids worry us, and we rejoiced in the swift rush. We swept past No. I Camp and saw the brush shelter that had been such a comfort to us on more than one occasion; it was hard to realise any conditions so rigorous today.

The last reach was passed and the last rapid run; we landed on the point that we had left buried under snow and ice and walked up to the house, perfectly stunned by the strangeness and difference of things. Everything looked so fresh and green; now that they were in leaf the house seemed fairly buried among the willows, and big blue flowers abounded through the woods, adding a vivid touch to the bright colours.

Limestone Walls of Canyons

The house looked curiously high; we cut the tin strips that sealed the door and found everything was just as we had left it. But the first impression of the interior, once so snug and attractive, was gloomy and dirty after the light and brightness we were used to of late.

We started a fire, swept out the dust and sand that had shaken down from the roof, and the place soon resumed its familiar air of snugness and comfort. I was in the midst of a much needed hot bath when the Doctor, Hornby, and Harry arrived with the dogs. We had a confused and disorderly supper; but it was good to be home again and to find our party happily together under this well–proved shelter after our long journey and varied experiences.

The next few days were busily spent in making preparations for our voyage across the lake.

Lion washed out the *Aldebaran* and varnished her; we had brought a tin of varnish for this express purpose. I had to develop all the pictures we had taken, then there was a great sorting out and selecting of the stuff we wanted to take with us. The load we could carry was strictly limited; it was heartbreaking to abandon some of our things, but there was no help for it; only such as were indispensable could be taken, everything else had to be ruthlessly left.

It was Hornby's intention to spend another winter in the country: many of the things, including some food supplies that were left, came in very useful to him.

Father Rouvier was to return in the summer with another priest to carry on the work of converting the Eskimos.

Perhaps it may be a pity that the latter cannot be left strictly alone; competent observers declare that civilisation means nothing but inevitable ruin and misery for them. But these deductions have been drawn from the fate of the Mackenzie River Eskimos where the conditions have been very different. At Coronation Gulf there is never likely to be a sudden inrush of civilisation in some of its lowest and least responsible forms, such as attended the booming of the whaling industry in the Beaufort Sea.

By their fruits ye shall know them. My own observations of the work carried on by the Oblate Fathers of the Mackenzie River district gives me an unbounded respect for these devoted self–sacrificing men. Their organisation is excellent, their methods matchless, their men well chosen and well trained for the work, their motives command admiration. The Indians' debt to them is inestimable; to them they owe all that gives grace, encouragement, and consolation to their lives.

We had already arranged with Father Rouvier that he could use our house the following winter, and he had undertaken to look after anything we left there, so we carefully packed the stuff that might yet come in useful and piled it at the end of the house.

The François family came on the last day, and we gave them many parting presents. They seemed quite affected at the prospect of our leaving,

Goodbye

more so than I would have imagined; they hoped we would come back some day soon.

We loaded the *Aldebaran* below the first rapids, where the *Jupiter* had been unloaded nearly a year before. We were gladdened by seeing old Jacob again just before we left, but the old fellow had come too late for the distribution of presents and we had nothing left to give him.

The last goodbyes were said to Hornby and our Indian friends; to the dogs that had served us so faithfully and well; and to the dear little house that had been such a sure home and happy haven. The last pictures were taken and on Wednesday, June 26th, we started on our long voyage across the lake.

We Begin Our Journey Across the Great Bear Lake

IX

HOMEWARD BOUND

## The Voyage of the "Aldebaran"

OUR voyage in the *Aldebaran* across Great Bear Lake was in some respects the most exciting part of our adventures in the North. But as our real work had begun with our arrival at the Dease River and the starting of our various tasks for the summer, so did it end when we returned to Hodgson's Point from our spring journey to the Coppermine. Only a general account can be given of a voyage that was longer and more hazardous than we had anticipated.

The *Aldebaran* was a big "freight" canoe made by the Peterborough Canoe Co.; she was 18'6" long, by 42" beam, and 18" deep amidship; built of basswood, in longitudinal strips, with close ribs. By the chances of construction that are familiar enough to any who have had much to do with small craft, she was rather more graceful in her lines than the ordinary big canoe built on that mould; we had remarked this on frequent occasions when direct comparison could be made.

She was rigged with a lug sail; the cloth we had got at Edmonton; Lion and I had made the sail on our journey down the river in the *Grahame*. It was amply large for the boat, but we were experienced in handling canoes under sail and could go to extremes in this respect; moreover we had provided it with reef points so we could reef down in a high wind. Those reef points often came in very useful, particularly the first two days; more than one occasion on this voyage tried our rigging and our skill to the limit, tried the mast especially which was made from a small fire–killed spruce that we had got near Smith's Landing. It was a very tough piece of wood; I have seen that mast bend like a whip.

When we left the Dease River we had a total load of over 900 lbs., excluding ourselves, whose combined weights must have added more than 450 lbs.

Considering this heavy load the *Aldebaran* travelled very well; her unusually fine lines gave her good speed. But those big canoes are in fact very hard to handle; they are most unmanageable in a wind and difficult to steer when heavily loaded.

Until we actually got out on Bear Lake we did not know in what condition the ice was. Such accounts as the Indians had given us were rather vague, but we hoped for the best and trusted to be able to get along somehow.

The bay behind Big Island was quite clear, but when we passed the straits beyond old Fort Confidence we were dismayed to find the surface of the lake covered as far as the eye could see; except for a small open space around the shore the ice lay intact as in winter.

For the first two days of our voyage we had a strong northeast wind; it kept the ice clear of the shore except at a few points where we had to shove though or carry and haul over. In spite of these delays at ice–jammed points we made good headway; on the evening of the second day, when the wind had dropped to a dead calm, we were close to the mouth of the Haldane River and had come over ninety miles from Hodgson's Point.

This part of the lake is inexpressibly dreary; the shores are low with long points of gravel and small boulders; the lake is shallow and there are many small low stony islands. No spruce trees grow within several miles of the lake and there was as yet no sign of reviving vegetation along this melancholy ice–bound northern shore; we had left summer behind us at the Dease River and were fated to see nothing more of it for nearly a month. The mosquitoes were the only things to remind us of the advancing season,

and these came in swarms, although the temperature was never above 42°
and occasionally down to the freezing point. We were obliged to wear mosquito veils and gloves again.

At a point some three or four miles west of the Haldane River our further progress was completely barred by the ice; it was jammed hard in
this part of the lake, the strong northeast wind had broken it up to a small

Detention Point

extent, and the floes were driven between the islands and the mainland;
out in the lake the ice was still quite solid. We were obliged to camp here
and wait for warm weather and a high northwest wind.

"Detention Point" was the name we gave our camp; it was near the
extremity of one of the long low gravel points, characteristic of this part of
the lake. There was a deep bay on either side of us; both bays were clear
of ice, but it was jammed hard along this and the next point. The bay to
the east of us, on the shore of which we had made our camp, was about
three-quarters of a mile across and more than a mile deep; at the end of it
was a wide sandy beach. The Haldane River came from the north to within a

mile or so of this beach, then it made a bend to the east and discharged into another bay beyond.

There were no spruce trees nearer than two or three miles, but we had a soft mossy place for our tents and plenty of driftwood for our fire.

Behind our camp was a low knoll of small boulders; it rose prob-

Camp at Detention Point

ably not more than fifteen feet above the level of the take, but it was the highest point for some miles around and served us as a good look-out.

It was the morning of June 27th when we made camp on Detention Point. Day after day went by, but the ice remained unchanged. The sun looked like a dull disc of silver set in the sky and seemed to have no more power. There was a kind of fine mist in the air during the day when the sun was high; at night time, when the sun got near the horizon, we usually had clear skies, and sometimes very beautiful clouds. We saw the sun at midnight for the last time at this camp; only part of its disc was visible over some mountains far to the north.

We spent the time fishing, hunting, sleeping, and eating; there were very few signs of caribou and those all old; they may have been made the previous

The Midnight Sun

autumn. There were a few ducks and geese, but our fishing was most pro-ductive. Trolling with a rather large spoon bait we caught lake trout as we wanted them; they were a most delicious fish; firm and fat, like all those fish of the large northern lakes where the temperature is near the freezing point all the year around. Those we got averaged about 11 lbs, with curi-ously little variation from this weight. The smallest we ever got weighed 7 $1/_2$ lbs., the largest 20 $1/_2$ lbs., but somewhere around 10 to 12 lbs. was the usual thing.

The fishing never failed us at any time we tried it; we always trolled in the bays, not off the points as Hanbury recommends; but of course the fish may have different habits at different times of the year.

The weather kept the same all the time; the days were cold, the sun was usually visible but with no more apparent power than the moon. The nights were usually clear, sometimes with a touch of frost.

Though the sun seemed so feeble and though the ice appeared so little altered from day to day it was in fact swiftly disintegrating and the chill and mist were caused by the rapid formation of water vapour from the melting ice. It was unapparent to us, but the good work was going on all the time.

On July 3d, after nearly a week of impatient waiting, a wind at last sprung up from the north and moved the ice out so that we could make our way along the shore again. We broke camp and started; after making about seven miles under sail the wind dropped and then came up so strong from the west that we could make no headway against it at all. It doesn't take a very strong head wind to stop one of these big canoes.

We were obliged to camp again, but towards evening the wind dropped and we could take to our paddles once more. The wind had opened up a channel along the shore nearly half a mile wide; we kept steadily at our paddles all that night, except when we stopped for lunch at midnight. This was on a very pretty point; the spruce came close to the lake here and we were chilly enough to appreciate a good big fire. We were all tired; it required an exertion of self–denial bordering on the sublime to leave that pleasant camping place and resume the weary job of paddling.

As we got farther west the open channel became narrower, occasionally we paddled among ice floes with new ice forming on the water between them; the shores became higher and more forbidding, the outlook more and more gloomy.

It was dead calm and the surface was covered with the small spikes and spars of freshly forming ice. Crossing one bay in particular we were much impeded by *slush* forming along the sides of the canoe just below the water line. It was a curious phenomenon and one that I am quite unable to account for. Before and since I have paddled over calm water on which ice was just forming but never saw this peculiar formation of slush on the boat except on this occasion. The conditions, whatever they were, that made the slush form lasted only about half an hour, although ice was forming for several hours and we passed through various areas of it in various stages of formation.

It was a serious check to us while it did last; it was not readily seen by the eye and we didn't realise what it was for a while until an accumulation near the bow made so much noise that Lion shoved his paddle along the water line to see what was the matter, and to our surprise scraped off the slush in masses like water–soaked snow. It extended abaft the beam and formed very quickly.

In my notes at the time I have written: "Was this in fact what happened to the *Fram* off the coast of Asia that Nansen attributed to and called 'Dead water'?"

Early in the morning we passed the eastern point of a deep bay; a couple of a islands are just inside this point; they are shown on the excellent map made by J. M. Bell, a map that we had with us and which we found helpful and accurate, at least as regards the part of the lake he had travelled over. Hanbury made the traverse across the lake from this point. He calls it Traverse Point and the bay Traverse Bay. He mentions the two islands near the point; we had expected to find something bigger. They are in fact mere gravel reefs.

We camped that morning on a high point to the west of Traverse Bay; it was a terribly bleak and desolate place, but the ice prevented any further progress. We had a good look over the lake from here, the highest ground we had been on for some time.

Smith's Bay is about twenty miles wide at this part; it was solid with ice and no traverse was possible for us till this broke up.

Our camp was a very disagreeable one; there was no driftwood any-where along the shore; some poles of a very ancient teepee that stood on a point a little farther on served us for fuel. We were all much fatigued; we had been travelling more or less for twenty–four hours and had paddled nearly forty miles. We didn't spend much time worrying over the appearance of our camp that morning, but turned in and had a good sleep till afternoon.

At 8 P.M. the ice had opened up enough to let us get a little farther ahead.

Sand Hills Camp and the Little Lake

We paddled for a few miles along that dismal gravelly shore, mostly threading our way among ice floes. Then we came to another large shallow bay with a prominent island off its western point. The bay was packed full of ice, broken up into floes of various sizes, and we had to camp again, but by this time the character of the shore had changed, and we were able to pick out a very snug and pretty spot for our tent among a charming little grove of spruce that grew in the sheltered corner of a high sandy hill overlooking the lake. It was a hard job to pack our stuff up that hill, but the camp we had on the summit made the work worth while. From our camp we had a fine view of the lake, though the prospect was not an encouraging one.

Sand Hills Camp we called it; behind our camp was a delightful little lake surrounded by spruce; why it should have been there at the top of a sandy hill we couldn't imagine, but there it was.

An Indian Grave

The country behind was nothing but short, steep, sandy hills, covered with plenty of moss and with fairly large spruce trees set at regular open intervals. Small lakes abounded among these hills; about a mile behind our camp was a larger one, a beautiful sheet of water very irregular in shape, with spruce–covered points, little islands, and curious bays.

Farther along the sandy hill was an old Indian grave; its palisades were much weathered and probably more than fifty years old. Around it the moss grew thick and soft, behind was the forest, in front the boundless lake, and over all the infinite unobstructed sky. One could scarcely wish for a more pleasant resting–place.

We spent several days at Sand Hills; the ice had us shut in so absolutely that we could not even fish, except from the shore near the mouth of a little stream where Lion caught some small blue fish with a rod and a little spinning bait. We hunted assiduously but got only two or three ptarmigan all the time we were there. It was a pleasant enough country to look at but very destitute of any animal life.

On July 8th, the ice opened out so that we could cross the bay. Sand Hills Camp was a comfortable one, and there was not much to be gained by going ahead as we were already at about the nearest point to make the tra-verse. However, anything was better than inaction so we loaded our canoe again and went farther west. We crossed the bay, passed the island at the western end of it, and rounded a headland of high gravel hills quite bare of any trees.

Beyond this headland a small river discharged into what appeared to be a deep narrow bay; westward of this bay a high, steep, sandy ridge runs parallel, and close to the lake, the immediate shore is low level turf–covered gravel with a fair amount of spruce along the ridge. The beach is stony, farther out the lake bottom is fine sand and clay.

We camped on the level space below the high sandy ridge where there was a good place for our tents. Here were some old signs of a former camp that we thought might have been made by Dease.

We were getting somewhat anxious now over the persistent continu-ance of the ice. Our steamer was due to leave Fort Norman in the begin-ning of August; if we failed to connect with this we would have to either track our boat up the Mackenzie River or descend to Fort Macpherson and attempt to get out by way of the Porcupine and Yukon rivers. Neither of these alternatives was a pleasant prospect to us. In 1837, Ritch of Dease and Simpson's expedition was delayed by ice at Gros Cap till the beginning of August. Towards the end of July we ourselves had seen the shore to the north of that cape lined with ice. In the second week of August, Ritch could hardly make the traverse from Acanyo Island. The ice looked so solid and the weather kept so cold that it seemed quite possible a similar experience might be in store for ourselves. Since our arrival at Sand Hills Camp the weather had been colder, with fresh ice forming among the floes almost every night.

The high ridge behind our camp gave us a good view over the lake. A lane of open water about a quarter of a mile wide extended along the shore as far as we could see to the west. The high land on the opposite side of Smith Bay was visible, but nothing could be seen of the ice conditions along the shore on that side. The prevailing winds had always been from some northern quarter, and we might find it a harder job to get along the south shore of the bay than the north. No map shows the end of Smith Bay correctly. Bell had crossed near this point so his valuable map was of no further use to us. In his report Bell says that the extent of the lake west— ward is much greater than hitherto supposed. After a long consultation we decided to go ahead and see what the end of the bay looked like, and next morning we made an early start.

It was a fine day, with a fresh north–northwest wind, but the shore usually afforded shelter and we were not hindered by ice. For the first eight or ten miles we found the coast of a pleasing character; the shore is generally low, often with small spruce coming close to the lake. Then came a high gravel point, barren and gloomy, with snow drifts in places right to the water's edge; the last snow we had seen had been on the Cop- permine Mountains. Everything about Smith Bay seemed to indicate se- vere conditions in winter and a long–delayed spring. Beyond this barren point there is a wide bay with a sandy beach and a delightfully pretty little river discharging into the lake. We called it Sleigh River from finding an old runner sleigh on the bank near it. To the west of this bay there is an- other bare gravel ridge; it rises perhaps a hundred feet above the surface of the lake and forms a prominent headland. From the summit of this point we obtained a good view of the end of the bay. The northwest corner is rounded and regular, the western shore appeared to be very high land also and regular in outline. In the southwest corner there appeared to be an entrance to another bay which extended as far as we could see from here. We thought it was about five or six miles across at the mouth; on its southeast was another high smoothly rounded hill.

The northwest corner of the bay was clear of ice, but we were not sure

Waiting

about the traverse across the bay in the southwest and the whole prospect looked so uncertain that we decided to return to our camp by the high ridge, and await the pleasure of the weather. We paddled back that same day while we still had open water; it was midnight when we got back to the old camping spot. We felt as though we had drawn back from some kind of a trap.

It proved an exceedingly lucky thing for us that we had turned back. The next day was dull and threatening with a strong east wind. We amused ourselves hunting ptarmigan and catching lake trout. During the night it rained heavily; we had put up both our tents; the Doctor was in one and

Lion and I in the other. At five in the morning I was wakened by the rattling of our pots blowing about. The long wished–for gale had come at last; it was blowing furiously from the north–west. Lucky indeed for us that we had stayed on this northern side of the bay!

I turned out to look around; the Doctor's tent was flat on the ground with the Doctor sleeping resignedly under the ruins. Lion and I had to build a brush windbreak at once to save our own from the same fate, yet we were not in an exposed place. The storm raged all day; by six in the evening there was not a vestige of any ice between us and Acanyo Island, though we could see it packed along the shores, both to the east and west in fields of floes.

The weather moderated and we made the traverse next day; a heavy swell was still running but the wind had fallen to a gentle breeze and we sailed most of the way across. I saw sandy bottom several time as we made the traverse; the lake was only eight or ten feet deep in places here. It was at this place that Bell had crossed Smith Bay; he calls the island L'Isle Sans Arbres; we always spoke of this as Acanyo Island, as two islands are shown on Franklin's chart in about this position, and we had supposed these were the same as L'Isle Sans Arbres. The chart I speak of as Franklin's was in fact made by Richardson, who was associated with Franklin on his second journey, and who spent part of the summer of 1826 in exploring Great Bear Lake between its outlet and the mouth of the Dease River. But there are several low sandy islands farther to the east, one of them several miles in length. These may be Richardson's Acanyo Island, though they do not lie in the same direction relative to each other.

The ice was broken up and packed in a dense field of floes along the coast to the east. We could get no farther, but were fortunately able to reach the mainland at the entrance of a deep bay behind L'Isle Sans Arbres, which Bell called Ice Bound Bay. We made a very comfortable camp about a mile inside the bay; we had got to a different looking country now, the plants and shrubs were in greater variety, many of them new and unfamiliar to us; the spruce trees were larger than anything we had seen for a year.

Ice Bound Bay is in fact a pretty sheet of water; what we saw of its eastern shore is low and densely forested near the water; behind these woods rise the high Eta–Tcho Mountains. The western shore is varied and hilly with the spruce more scant and sparse.

Again we were reduced to watchful waiting, we put in the time fishing and hunting; bears' signs were plentiful and recent, but we saw no bears; the fishing as usual was more profitable, so we laid in a good supply of trout while we had the chance; the ice might close in at any time and prevent us from fishing. We trolled in a little bay near the point; every time we made a turn around that bay we caught a trout, until we had enough to keep us for several days. They weighed 7 $1/2$, 8, 9, 9, 9, 10, 10, 10, 10 $1/2$, 11, 11 $1/2$, lbs. I give the weights as they show how very uniform in size were these fish that we caught.

Our stay at Ice Bound Bay was shorter than we had dared to hope. On the evening of the day after making the traverse a gentle breeze sprang up from the south, and the ice began to move out in the lake, soon leaving a free passage for us along the shore. We struck camp at once and made about ten miles that night; the fair weather held next day and we continued our voyage until evening; paddling all day long against an incessant troubled, swell, though there was not a breath of wind, and the sun was brighter and warmer than we had seen it since leaving the Dease River.

The ice still kept moving away from the shore, though there had been very little wind; at the latter part of the day we could see only its brilliant white line far out on the lake. In the evening we came to a fresh barrier of broken ice that had evidently been packed in a large bay on the northern side of the peninsula. The shore was lined with floes to a depth of a hundred yards or more, and a field of them shaped like a long tongue stretched across the bay.

While coasting along the edge of the ice we saw a big brown bear gravely walking around on the top of a hill close to the shore; it was four or five hundred yards away, too far to shoot from a canoe rocking in a swell. We managed to shove through the floes and to land, but by that time the bear had found business elsewhere, and we saw no more of him. He was an immensely powerful brute; seen through glasses his big shoulders and enormous quarters had something almost elephantine about them.

We had to make a long détour out in the bay to find a place in the ice-field where the floes were open enough to allow us to pass, but we were able to shove through, and camped on a bleak point at the western side of the bay; it was too late to attempt the traverse that night. The beach was lined with big boulders that made landing a rather risky job in the swell.

It blew from the northeast next day; the ice closed in on the point and the floes ground against the boulders in a heavy swell. We were held up till evening much against our will; we were losing a fair wind and there was something particularly unpleasant in the location of our camp. We had a fairly good spot for our tent, but the exposed situation and those big boulders that lined the shore made a very disagreeable impression. The swell moderated by evening; we managed with some risk and difficulty to launch and load our canoe over the grinding floes, and at 9 P.M. we started under sail.

The wind increased again; it is about twelve miles across the bay and we made the traverse in only a little over two hours; in spite of her heavy load the *Aldebaran* ran swiftly and gracefully through the constantly rising sea. As we rounded the coast leading to Gros Cap the wind got stronger and ever stronger, but we had a windward shore and at least the water was quiet.

We scudded across the well–remembered bay where fogs and calms had held us up in the *Jupiter;* we were reefed down now with our mast bending like a reed and the boat buried to the gunwales in foam. The strong wind lasted till we rounded Gros Cap; then it died away very suddenly, and we took to our paddles until we got to the best place to begin the traverse across Richardson Bay. We were well satisfied with our night's run; we had come nearly thirty–five miles in eight hours.

On the morning of July 17th we made the traverse, and our voyage henceforth was unmitigated pleasure. All the ice had been left behind us; the end of our journey was in easy sight; there was nothing to worry us, and we could, give ourselves up to complete enjoyment of a delightful canoe voyage on beautiful waters along an interesting coast. The weather kept fine and tranquil; the latter part of our voyage was a succession of halcyon days; indeed we were in a humour to appreciate and enjoy them after the experiences of the last three weeks.

On the evening of July 19th we arrived at the little Indian settlement at the end of the lake. To our surprise the place was completely deserted; not a soul was there. The Indians had not yet come up the river from Fort Norman. As they come back to the lake soon after the arrival of the first steamer we supposed something might have happened to delay her, and began to entertain hopes that we might even yet catch this boat on her first south-bound trip.

We paddled across to Little Lake, intending to camp that night at our old place where we had refitted the *Jupiter*. As we entered the narrow channel to the lake we saw some Indians camped farther along the shore. We went to interview them; they were some of the women folk of the Bear Lakers who had gone to Fort Norman, and could only tell us that the Indians had not come back and they did not know when to expect them.

We had tea at our old camp site; no one had been there since our last visit in the *Jupiter;* odds and ends of our former stay lay around as though we had never been away at all; here were a pair of shoes that had begun a protracted career of useful service by tracking up a river a far way from the Great Bear Lake and had finally worn out on their long hard walk up the banks of the Bear River; the stakes we had used for our mosquito bars were still standing, and we lit our fire with the chips made when converting the *Jupiter's* steering sweep into a mast.

Our tea was a somewhat solemn one, yet cheerful withal; the place recalled vividly our thoughts and general frame of mind when we had camped here last, more than a year before. Our adventures had then been all before us, and we could look back with satisfaction and pleasure on the good fortune that invariably had been our lot.

The more we thought over the non–arrival of the Indians the more likely it seemed to us that we might yet catch the steamer. The Indians make the trip up Bear River in a week; the steamer takes about eight days on her round trip to the posts below Fort Norman. Supposing that the Indians had left within the same time of the steamer's arrival as they had done the previous year our chance of catching her was a very good one.

Although it was then nine at night and we were all pretty tired after a long day's journey, we decided to continue on our voyage until either we met the Indians on their way up the river, or until we got to Fort Norman.

We entered the Bear River at ten, taking almost with regret a last view of that great inland sea that had been so welcome a sight twelve months before and that had on the whole treated us in such a friendly way.

The swift current of the Bear River seized on the *Aldebaran* as a mere plaything and hurried us down at a bewildering speed. The night had become overcast, and it was dusk enough to make steering down some of the swifter places a matter of risk and difficulty. But we felt that the time had come to take chances; we imagined ourselves arriving at Fort Norman just in time to see the steamer pass out of sight to the south; every minute was of consequence to us now.

About ten miles below the lake we passed some Indians camped on the river bank. They were on their way from Fort Norman; none of them understood any English or French, but we made out that Father Rouvier and the rest of the Indians were on their way up the river a day's journey behind, and that the steamer had sailed on Sunday downstream. It was now Friday; they meant Sunday a week before the last, but we didn't know that at the time; their information confirmed us in our hopes and we kept on our reckless course through the uncertain light.

At two in the morning we arrived at a large camp; the whole outfit was here; Father Rouvier, Father Le Roy, Johnny Sanderson, and Jimmie Soldat, and the rest of the Indians.

We roused Johnny and turned out Father Rouvier; our first question was about the steamer, and we learned to our grief that she had sailed *south* the Monday before. Our hopes had been very lively, but they had originated only a few hours before and we soon got over our disappointment.

Bear River near Bear Lake

Father Rouvier was making his way back to the Dease River; he had a companion priest with him, Father Le Roy, and the Indian Jimmie Soldat.

It was no use going any farther that night; we put up our tent and the Fathers came around and had a good feed of pounded meat and grease. We still had a fair supply of this favourite dish of the North, but it was a treat to the Fathers, who had been strangers to it for some months.

We told Father Rouvier of our doings since he had said goodbye to us four months before, and in turn he told us the latest accounts of the great world. The loss of the *Titanic* was the news which made the greatest impression on us. It was after four in the morning when we turned in, and our untimely arrival and disturbance had so upset the camp generally that they did not get started till nearly ten that day.

Indians Tracking Up–stream

The morning was bright and warm; it seemed oppressively hot to us. Judging by our senses alone I should have *guessed* a temperature of certainly over 80°, perhaps as much as 90°, in the shade. A sling thermometer recorded 62° which shows how little mere sensation may be relied on; we had become so accustomed to the uniform cold weather on the lake that we were quite incapable of judging.

We saw the outfit get started, thankful now at any rate that we were *not* travelling in the same direction. They had nothing but canoes; the

Fathers and Jimmie Soldat had a new big canoe with a heavy load. Hornby had been hoping that they could bring him in some supplies; they had been quite unable to do this and we were glad to think that he was well fixed for several months at least with what we had been able to leave him.

The Fathers on their journey up the Bear River

As soon as the others had started, we struck camp ourselves and continued our journey down the river. It was a different thing steering in full daylight to what it had been in the doubtful twilight of the night before, and the swift motion was delightful.

We ran the rapids in triumph; they are really safe and easy enough to run, though so hard to ascend, especially with a York boat; the difficulties are due mostly to the perilous character of the shores. We realised here that the water was much lower this year than it had been the previous one, and the ice cliffs were not nearly so formidable now as they had been on our journey up. The Franklin Mountains looked very fine and we were in a much more favourable mood to admire them than when we had seen them last.

We camped early that afternoon; we did not want to arrive at Fort Norman late in the evening and time was of no object to us now. We had

Mount Charles—From the East

come back to regular summer again; after being strangers to them for a long while we saw poplar and birch, beautiful in their full pride of early summer; the size and variety of the trees were quite bewildering to us after the invariable small spruce that we had become so accustomed to.

Resuming our journey next day, we floated swiftly down the lovely lower reaches of the river; Bear Rock rose higher every minute and soon the great Mackenzie River came in sight and the snowy summits of the far distant Rocky Mountains.

At two in the afternoon we left the clear waters of the Bear River behind

us a last goodbye to an old friend; muddy waters were to be our highway
henceforth for many weeks to come.

I am bound to say that our arrival at Fort Norman was very flat; we
had pictured the hearty welcome and general interest that are the conven-
tional reward of travellers when they arrive at the first remote outposts

Our Camp at Fort Norman in July, 1912

of civilisation. The first people we saw were two priests taking a Sunday
walk along the shore; our arrival was a matter of complete indifference to
them; they did not even return our greeting, but simply turned their backs
and continued their stroll. Why the very birch and poplar trees had given
us a better welcome than this!

We landed below the place where Hornby and Melville had been camped;
some of the other inhabitants of the post now showed up, Hodgson and his
son; Fair, the factor of the Northern Trading Co., and an old trapper called
Store. They gave us a kindly hand to land our stuff and to carry it up the

steep bank to the place where we had decided to locate our camp. We put the tents up and had everything shipshape before seeing any more of Fort Norman.

How woefully disappointed we were in our welcome we have seen; the first inhabitants we met at this post had greeted us by turning their backs.

Family Birch Dark Canoe

Worse, far worse, was to follow. Instead of being hospitably entertained at the factor's house, it was the factor who came to our camp and was hospitably entertained by us; we could even produce a bottle of cognac to celebrate the occasion!

It is true that we had failed to conform to the convention requiring that the explorer should come to the first post ragged and half-starved, eating his moccasins and mits. So it was really all our fault in both cases, and we only got the proper punishment that in some shape or other is inevitably meted out to all offenders against convention. And our good friend, Mr.

Leon Gaudet, was as heartily welcome to us as doubtless we would have been to him had our condition been properly miserable.

Our stay at Fort Norman proved much longer than we had anticipated. The first fortnight after our arrival passed pleasantly enough; the weather was invariably fine and warm, our camp was comfortably situated and

Roman Catholic Mission

commanded a lovely view; the last four months had been fairly strenuous work and the rest was welcome enough. Then we had a chance to observe the life at a fur trading post under its usual conditions instead of at the times of abnormal activity attending the arrival of the steamer, the only occasions on which we had seen them hitherto.

Little enough indeed went on; our own life at Hodgson Point in winter time was scarcely less devoid of incident. We became better acquainted with Father Ducot of the Mission, and with Mr. Hodgson, two men of great experience in the North who had many interesting things to tell us.

The Fathers showed us around their church with justifiable pride; it was profusely decorated inside with paintings on the wood of the structure, the result of incredible care and pains.

Once in a while a party of Indians would come in and pitch their tepee on the beach, and after getting a few things in trade, would go off to hunt again.

Indian Camp at Fort Norman

The trading store was like a small, rather badly stocked country store. The quality of the goods carried was generally excellent, a feature of the Hudson Bay Co.'s fur trading stores that deserves the highest praise. Everything was very expensive but not unduly so, considering the difficulties and distance of transport, and the high quality of the goods; in this respect Fort Norman would probably compare very favourably with the towns on the Yukon.

When trading they allowed only one Indian in the store at a time; if more than one was allowed to enter the others would give so much advice

that the negotiations would never come to a conclusion. Their behaviour in the store was exactly like that of uncertain children. I
watched an Indian select six "skins" worth of stuff one day. He was
fairly dazzled and confused with such wealth to choose from. He
took a little tea and a little tobacco, then came a long mental strug-

The Skin Boat

gle between a black ostrich feather and a blue leather peaked cap.
The ostrich feather finally carried the day, and he took it away in
triumph. Truly "the first spiritual want of a barbarous man is decora-
tion, as indeed we still see among the barbarous classes in civilised
countries."

A party of Montagnais Indians turned up at the post one day from some
far eastern tributary of the Mackenzie. They had killed a number of moose
and with the skins for cover and saplings for a frame they had built a large
boat and journeyed down the river bringing in the meat. Some of it was

dry, and some fresh, or green would be a better term, unless for choice and truth one said simply rotten.

The boat showed great ingenuity in design and construction, and was really a thoroughly serviceable craft.

As soon as they had disposed of their meat, they crossed the river and struck by some overland trail back to their hunting grounds on the eastern slope of the Rockies. They sold their boat to the factor of the Northern Trading Co. who broke it up for the skins.

The *Mackenzie River* was due to arrive on August 1st; the date came but not the steamer; the days passed and still there was no sign of her.

Our stay at Fort Norman had hitherto been agreeable enough, but now ensued a period of anxiety and uncertainty; there was no way of telling what had happened to delay the boat nor even whether she could come at all.

We thought a good deal of attempting to get out by way of Yukon, but this was a long trip and not to be entered on lightly at that time of the year. If we had been reasonably certain of getting help at Fort McPherson to carry our stuff over the eighty–mile portage to the Porcupine River we would not have hesitated; but if we failed to get this help, as we well might, our condition would be worse than ever.

We had arrived at the beginning of summer; we saw things come to maturity with astonishing swiftness. Then came the pause of consummation; Time had completed his gifts and was soon to confound them.

The nights got darker and colder; the dusk beginning to close in about ten in the evening affected us most as a sign of the rapidly passing season.

It was part of the day's routine to stroll up the river shore in the afternoon, the time at which the steamer was most likely to arrive. A short walk above the post took us to a point from where we had a view of ten or twelve miles up the river. We would sit on the big driftwood logs thrown up in great numbers by the high spring water, whittling the soft dry cottonwood sticks and hoping every time we looked up that we might see the steamer rounding the point in the distance. Any moment might bring her, but day after day passed in continual expectation and continual disappointment.

At last we decided that if the steamer did not come by August 21st, we would start for Fort McPherson and take a chance on being able to cross the divide and reach the Yukon before the steamers on that river were laid up for the winter.

# X

## STEAMER AND SCOW AGAIN

O N August 17th, we were just getting ready for tea at our camp, when there was a great shouting at the post and we saw the flags go up at the various flagstaffs. Our long wait had come to an end; the steamer had arrived at last; there she was rounding the point above the post where we had so often watched for her in vain; she seemed strangely small out on that great stream.

Our old friend, Captain Mills, was not on board; he had been engaged in building another steamer on the upper Athabasca. But we had the pleasure of meeting Mr. Pearce again; and Captain Mills's successor, Captain King, well sustained the high standard of courtesy that we had been used to in all our dealings with the Hudson Bay Co.

We broke our long-settled camp with few regrets; the *Aldebaran* was loaded up again and we were soon settled on board the *Mackenzie River* once more.

A long and weary road still lay before us; two months were yet to elapse before we got to the end of our journey; two months of slow progress and hard struggle against the river that had brought us down so swiftly and so easily.

We had seen the turn of the season while waiting for the steamer; now we saw the days shorten with redoubled speed as the sun receded and as each day found us in a lower latitude.

We left Fort Norman on the morning of August 18th and on the evening of August 29th we arrived at Fort Smith.

The *Grahame* had gone south several weeks before; fortunately we had brought our old reliable *Aldebaran* with us on the steamer and were still in plenty of time to connect with a scow that the Hudson Bay Co. were sending up the river from Fort Chipewyan.

We had all our stuff, including the canoe, sent over by wagon to Smith's Landing, and walked there ourselves; within twenty–four hours of the time we had arrived at Fort Smith, the *Aldebaran* was once more loaded up and we were under way again in our own craft. It was her last voyage and in some respects the most pleasant.

The Slave River above Smith's Landing is swift, but the rocky points make so many eddies that we could take advantage of these and so paddled most of the way.

The weather was delightful, warm days with cold nights and misty mornings, regular early Canadian autumn at its best. This part of the journey is very beautiful; there is more variety in the trees and constant charm; there were no flies and we had plenty of time; altogether the last voyage of the *Aldebaran* was to us one of unalloyed pleasure.

We reached Fort Chipewyan on September 3d; the lake was much lower, and what on our way down had been open water between the post and the entrance of the Rocher River was now great level plains of mud with the river winding among them, and the rocky islands looking strangely out of place in a desert of alluvial silt. The *Grahame* was there, about to be laid up for the winter; we took up our quarters on board with the Hudson Bay Co.'s officers who were going out on the scow, and there we awaited the arrival from Fond–du–lac of the tug *Primrose* that would tow the scow as far as Fort Mackay.

The *Primrose* arrived from the end of the lake on September 10th; the *Grahame* was moored for the winter in a secluded "sny"[1] and on September 13th we started on the last lap of a long journey, the longest at least in point of time, and the part of it that was to impose a greater strain on our patience than anything that had happened to us yet.

[1] Corruption of *chenal*.

Besides ourselves there were as passengers a corporal of the R.N.W.M.P. and a crazy Indian woman under his charge; and Duncan MacDonald, a member of the Indian treaty–paying party. The Indian woman had been brought down the Liard River by Duncan; they came up from Fort Simpson with us on the *Mackenzie River,* and on arriving at Fort Smith the Woman

The Hudson Bay Co.'s Party and Their Passengers

was turned over to the police who sent her out to Athabasca Landing under the charge of Corporal La Nauze. No one understood her language; she came from some distant tribe on the eastern slope of the Rockies and her talk was as unintelligible to the Crees and Chipewyans of our crew as it was to ourselves. We called her "Matilda," her madness was at least of a cheerful inoffensive nature; she would laugh and sing and dance by the hour, and altogether she added very much to the gaiety of the crowd.

The first part of the journey to Fort Mackay was pleasant enough with the *Primrose* to tow us; we made good headway and she gave us protection and warmth on cold days. The season was advancing with rapid strides, the autumn colours were now lovely; sometimes there would be large areas where pink tints prevailed, but the predominant tone was pale yellow; overwhelming masses of uniform yellow would alternate with the unchanging sombre spruce.

Fort MacKay

We reached Fort Mackay on September 18th and camped a little above the post on the opposite shore. Here we hauled the *Primrose* out of the water, laying her up for the winter.

Then we took to the tracking line, but it was not till we struck the rapids above Fort MacMurray that our troubles really began.

The scow was a small one; it was heavily loaded for upstream work, and the tracking crew was short-handed. Following the long-established custom of the country, the general charge of the job was in the hands of the

Indian pilot. In this case he was a good man and a very hard worker himself, but he had no authority over the crew and no idea how to handle a crowd of men. The Hudson Day Co.'s men never interfered in any way and the outfit moved or stopped at the sweet will of the tracking crew.

The *Primrose*

We had left our canoe at Fort MacMurray, and we regretted at the time not having tracked it up ourselves. But between MacMurray and the Pelican Portage, a distance of 150 miles, the Athabasca River is in fact a very difficult and dangerous river to track up with a canoe. It was not till we were past the bad places that we were in the proper frame of mind to judge the performance of the crew, to give them just credit for their really hard work, and to realise how much wiser we had been in staying with the slow but safer scow. At the time the incessant delays and slow progress were intensely irritating to us, who had been accustomed to doing things in a much smarter way and to whom it was thoroughly humiliating to be so completely dependent on the pleasure of an unorganised bunch of hand–to–mouth half–breeds. We were all impatient to get home, and when we arrived at Fort MacMurray we had thought we were pretty near our journey's end.

The following, copied from my notes made at the time, a day taken at

A Long Hard Struggle

hazard, will give some idea of the difficulties and slowness of our progress, and of our impatient temper.

"*Tuesday, Sept. 24th*: Cold this morning; we had several degrees of frost last night and the mud near the water frozen. All hands had breakfast, then we hauled the scow a little farther up; there is a small cascade and we lightened the scow just below it. Turning a fine day, leaves beginning to fall fast and only here and there a touch of green to be seen anywhere.

"Carried the stuff up about a quarter of a mile, loaded up and made a start again at 10 A.M. Stopped 10.15 to change heavy line for a light one;

we were then at a comparatively quiet stretch of the river. The men had dinner here and we made a start again at 10.52. Stuck at 11, on again 11.15, stuck again at once, and fooled around till 12. We crossed the river then and had lunch on the other side. Water very shallow along here, constantly sticking and making scarcely any headway at all. From 1 till 2.30 nothing

Slow but Sure

doing, then the men had tea." (I hope the irony of this will not pass unnoticed!)

"Made a start again at 3.00 and attempted to go outside a shallow bar; the line broke and the scow drifted down to where it had been before tea; a hell of an outfit this is, unspeakably sloppy, the scow too small, no tracking line worth a damn, not enough men, and no one running things. We finally got up past Meatsu Point, then came a quiet stretch of the river for about a mile, then shallow water, and we stuck again. After fooling around for some time they decide to camp. . . ."

It took us seventeen days to reach the Pelican Portage, an average of less than nine miles per day. The weather turned cold and colder, the leaves all fell, the banks of the river were frozen, and it looked as though we might be hindered yet by ice forming on the stream.

At last we passed the Pelican Rapids and on the quieter water above we made better time; with the end of the journey in easy sight every one worked with a will; we took spells on the line, and at this quiet part of the river we could rest on the scow between tracking spells.

Then the weather turned fine and warm again; it was the Indian summer of the year, with nature in its most enchanting mood. My last memories are pleasant ones only: of quiet waters and comfortable camps, of fine nights and finer days, of short spells of work and long spells of rest; sitting at ease in the scow, lazily watching, through an atmosphere of uncomparable purity and ineffable calm, the naked trees reflected in the tranquil stream in all their beauty of line, and the faint silver threads of gossamer floating in the still air. Time itself had come to a standstill; such afternoons seemed as though they might last forever. Eighteen months before the North had welcomed us in all the ecstasy of spring; now that the time had come to bid farewell it showed itself in another and even more alluring mood, determined if it could not hold us that our crowning recollections at least should be of intimate charm and poignant beauty.

We saw the houses of Athabasca Landing come into view again as we had watched them disappear so long before. A few minutes more we were alongside, and our northern adventure had come to an end.

# XI

## CONCLUSION

A T first there was no strangeness in getting back to civilisation. The contrast was so complete and the interests so different, and yet so familiar that we took up our civilised life where it had been broken off ; we resumed at once our old set of ideas and our old ways of doing things. Some time passed before we began to feel in many subtle ways the results of a long absence. In regard to the great world we were the same people who had left it eighteen months before, but while we were relatively unaltered our world had gone on its appointed course, and unhastening, unceasing, the appointed changes had been wrought. It was ground irrevocably lost; no skill, nor energy, nor address, could recover it. The times had changed, the change in ourselves had no reference to them but made conformity to established usages more than ever difficult. You get nothing for nothing; for everything in this world a price is exacted. God is *not* given away, nor can heaven be had for the asking.

Two more views of us, and I have done.

The first a few days later. We are passing rapidly eastward over the great Canadian plains, sitting in warmth and comfort in a luxurious dining car. No slimy, muddy moccasins are on our feet now, and no caribou hairs on these steaks, but we didn't think the bread any better than our bannock, and the soup is distinctly inferior to that we made at Hodgson's Point.

Without it is snowing hard and the windows are so blurred that only a vague haze of driving snow and swiftly passing telegraph poles and fence post can be seen.

We are talking of our late companions of the north: of La Nauze and the Indians making their way back to Chipewyan in such weather as this; of Hornby and how winter must have set in with him. We regretted more than ever that he was not with us, and we wondered whether he would have revelled in the bill of fare as much as he always said he would.

A few days later and the last scene.

The Doctor has left us now to meet his own friends. Lion and I are together once more in a canoe; the *Alouette* seems strangely light and swift after the heavy *Aldebaran*. For a load we have only ourselves and the small bag containing our records, the bag that has come so far and by such varied means. We are finishing our long journey as we started it; we are back on clear waters and among woods of unequalled beauty. In our flying trip across the continent we have overtaken autumn; even yet some trees show scant remnants of their summer green. That dear poplar–crested ridge comes in sight; the trees have held their leaves to the last to welcome us, and red against their yellow, the flag proclaims that we are home at last.

# L'ENVOI

Flights of returning fowl that fling
In changing runes across the blue,
What time, incredible in hue,
    The Barrens flush to Spring.


Dark herds of hurrying deer. The call
Of beast and bird and south wind rains.
Deep whisperings in the bleached moraines
    Where stones and water fall.


So comes brief Summer to your lands
Lady of Silences serene.
How doth it profit us, O Queen,
    The labours of our hands?


Naught. We have come, and seen, and known
Full franchise—only to depart
Leaving, as hostage of each heart,
    All that was most our own.


You hold it now. No Spring may wake,
But, midst its alien imagery
We see blue tracts of rippling sky
    Invade the iron lake.

No skyward–lifting smoke uncurled
But straight there glows, in acrid haze,
Some fire of grey beleaguered days
    Whose light defined our world.


No wind that sweeps embattled spears
Of pine, no brawling stream at play—
But shouting rapids rise in spray
    To stun our deafened ears.


While smoke shall rise, while wood shall burn,
While stars shall wheel and waters run,
Thou weavest with the wind and sun
    Charms that constrain return.

    .     .     .     .     .     .     .


What is our count? The best we know?
The crown of Effort—all Desire?
A broken camp, a burnt–out fire,
    A sledge–track in the snow.


            BRYCE McMASTER, ("Clansman.")

# APPENDIX A

[From Franklin's *Narrative of a Journey to the Shores of the Polar Sea, 1819-20-21-22*. Geognostical Observations by John Richardson, M.D.]

THE Copper Mountains consist principally of trap rocks which seem to be imposed upon the new red sandstone or the floetz limestone which covers it. A short way below the influx of the Mouse, the Coppermine River washes the base of some bluish–grey clay–stone cliffs, having a somewhat slaty structure, dipping to the north at an angle of 20°.

The Copper Mountains appear to form a range running southeast and northwest. The great mass of rock in the mountains seems to consist of felspar in various conditions; sometimes in the form of felspar rock or claystone, sometimes coloured by hornblende, and approaching to greenstone, but most generally in the form of dark reddish–brown amygdaloid. The amygdaloidal masses contained in the amygdaloid are either entirely pistacite, or pistacite enclosing calc–spar. Scales of native copper are very generally disseminated through this rock, through a species of trap tuff which nearly resembled it, and also through a reddish sandstone on which it appears to rest. When the felspar assumed the appearance of a slaty claystone, which it did towards the base of the mountains on the banks of the river, we observed no copper in it. The rough, and in general rounded and more elevated, parts of the mountain, are composed of the amygdaloid; but between the eminences there occur many narrow and deep valleys, which are bounded by perpendicular mural precipices of greenstone. It is in these valleys, amongst the loose soil, that the Indians search for copper. Amongst the specimens we picked up in these valleys were plates of native copper; masses of pistacite containing native copper; of trap rock with associated native copper, green malachite, copper glance or variegated copper ore and iron–shot copper green, of greenish–grey prehnite in trap (the trap is felspar

275

deeply coloured with hornblende), with disseminated native copper; the copper, in some specimens, was crystallised in rhomboidal dodecahedrons. We also found some large tabular fragments, evidently portions of a vein consisting of prehnite, associated with calcareous spar and native copper. The Indians dig wherever they observe the prehnite lying on the soil, experience having taught them that the largest pieces of copper are found associated with it. We did not observe the vein in its original repository, nor does it appear that the Indians have found it, but judging from the specimens just mentioned, it most probably traverses felspathose trap. We also picked up some fragments of a greenish–grey coloured rock, apparently sandstone, with disseminated variegated copper ore and copper glance; likewise rhomboidal fragments of white calcareous spar, and some rock crystals. The Indians report that they have found copper in every part of this range, which they have examined for thirty or forty miles to the northwest, and that the Esquimaux come hither to search for that metal. We afterwards found some ice chisels in possession of the latter people twelve or fourteen inches long and half an inch in diameter, formed of pure copper.

To the northward of the Copper Mountains, at the distance of ten miles, in a direct line, a similar range of trap hills occurs, having, however, less altitude. The intermediate country is uneven, but not hilly, and consists of a deep sandy soil, which, when cut through by the rivulets, discloses extensive beds of light brownish–red sandstone, which appears to belong to the new red sandstone formation. The same rock having a thin slaty structure, and dipping to the northward, forms perpendicular walls to the river, whose beds lie a hundred and fifty feet below the level of the plain. The eminences in the plain are well clothed with grass and free from the large loose stones so common on the Barren Grounds, but the ridges of trap are nearly destitute of vegetation.

Beyond the last mentioned trap range, which is about twenty miles from the sea, the country becomes still more level, the same kind of sandstone continuing as a subsoil. The plains nourish only a coarse short grass, and the trees which had latterly dwindled to small clumps, growing only on low points on the edge of the river under shelter of the bank, entirely disappear. A few ranges of trap hills intersect this plain also, but they have much less elevation than those we passed higher up the stream.

# APPENDIX B

# REPORT ON A RECONNAISSANCE ALONG THE LOWER COPPER-
# MINE RIVER

## BY AUGUST SANDBERG, PH.D.

## TOPOGRAPHIC FEATURES

No high mountains exist in the area here considered, which extends from Dismal Lake eastward to a few miles beyond the 116th meridian, and trends northward approximately following this meridian from 67th parallel to the mouth of Coppermine River.

The Copper Mountains, by which term the high land trending east–southeast from Dismal Lake to Coppermine River and beyond, is designated, are formed by a series of basalt ridges with the same general trend as the range, and occupy a belt about fifteen miles wide. Towards the south they terminate abruptly in a nearly straight line, for miles, dropping with a perpendicular wall to the broad valley of slight relief adjoining the mountains to the south. The mountains attain only an elevation of 1200 to 1500 feet, presenting the appearance of a plateau, interrupted by a number of mutilated ridges, facing south with perpendicular cliffs of varying height, and sloping gently towards the north.

The Coppermine River, traversing the valley, with a northerly course to the south of the mountain, enters the Copper Mountains about five miles below Kendall River. Striking the hard basaltic rock it curves eastward and assumes a course practically parallel with the trend of the basalt ridges for a distance of twenty miles, before it cuts its way through the ridges with a curve toward the northwest and finally emerges with a northerly course on the plain to which the

Copper Mountains slope towards the north. In its passage through the double curve the river has cut deep, and in some places has made a narrow valley through

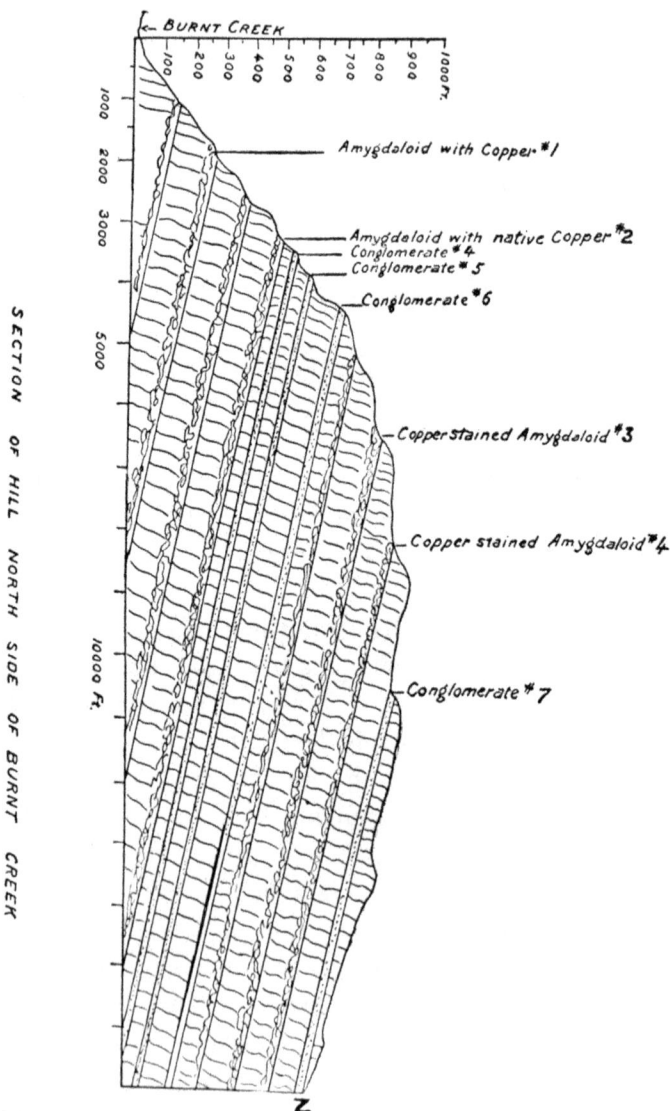

the mountain. A number of small creeks flow at right angles into the river and drain the mountains through narrow, constricted gulches in many places dividing the hills in detached blocks. From the valley the mountains rise by steps in the nature of terraces to the summit. The highest altitude is attained by three

SECTION OF BASALT RIDGE NORTH OF SANDSTONE RAPID.
(Vertical Scale Five Times Horizontal)

FIG. 2

adjoining peaks at the point where the river begins its eastward course. The bottom of the valley in many places is occupied by low ridges and small hills consisting of clay and gravel of glacial origin. On the northwest course of the river where the valley is quite broad, these clay and gravel deposits extend about a mile from the river on the south shore forming a typical miniature glacial land scape. A good growth of spruce is sustained from this soil, especially along the north shore of the east course and on both sides of the northwest course of the river.

To the north of Copper Mountains the country presents the character of a plain with slight relief, traversed with narrow basalt ridges of the same general trend as the Copper Mountains. Only the first and last of these ridges attain an elevation of about four hundred feet. Through this plain the river has cut a channel about one hundred feet deep with perpendicular sides of sandstone, alternating in some places with shelving clay banks. Where the river cuts through the basalt ridges the channel becomes tortuous and constricted to about fifty yards and less from its average width of about three hundred yards.

GEOLOGIC SKETCH

The results of our observations regarding the distributions of the geological formations are represented on the accompanying geological map. In making the map the observations of latitude and longitude made by Franklin have been used. All locations represented on the map have been made with compass and pacing. Magnetic declination 48° East.

## LIMESTONE

Along the southern edge of the Copper Mountains limestone forms part of the cliffs, with which the mountains frequently terminate toward south. On a fresh fracture surface the limestone shows a highly crystalline texture, generally fine to medium grained, of impure white or grey colour, which sometimes has a reddish tinge. Farther south between the mountains and Kendall River the in part precipitous west shore of Coppermine River is composed of limestone cliffs about one hundred feet high. At the exposure in the gorge, through which Kendall River flows into the Coppermine, the limestone beds are interleaved with thin strata of fine–grained red sandstone. The dip of the limestone is flat towards the north under the mountains.

## COPPERMINE SERIES

The Copper Mountains are composed of a series of superimposed flows of basaltic lava, which occupy a belt about sixteen miles wide in a direction at right angles to their strike. Their lateral extension is large, probably reaching east more than two hundred miles across Bathurst Inlet, where Hanbury describes the occurrence of basalt with native copper. Westward, and about forty miles north of great Bear Lake, there is said to exist basalt, exhibiting the same general character as the basalt at Coppermine River. Interstratified with the basalt are a number of detrital beds of reddish conglomerate. These occupy various horizons, but as far as our observation goes, they are more numerous in the upper part of the series. The basalt occurs in distinct beds of varying thickness, striking approximately parallel with the range. Westwards at Dismal Lake the beds strike east 22° south and show a dip of only 8° towards the north. At the great bend of Coppermine River the strike is approximately east 12° south and in the upper part of the series at Burnt Creek, the beds strike a few degrees north of east. The most common dip is 12° north. The effect of the bedded basic rock upon the topography is everywhere marked; a steep cliff facing south and a long back–

ward slope towards the north forming a shallow drift–covered valley between the crest of one and the rising cliff of the succeeding bed. The small streams draining the mountains into Coppermine River divide the hills in detached blocks, rising terrace–shaped from the bottom of the valley.

Some thick beds show columnar jointing in their exposed cliffs, but more often the jointing has broken the beds into irregular blocks with the jointing surfaces dipping, steeply towards the north. Usually the beds present a twofold division, an upper narrower amygdaloidal and a lower compact nonamygdaloidat portion. The lower massive part of the bed is dense, crystalline, medium to fine grained, of dark grey nearly black colour, which in some places changes to reddish brown. The compact and amygdaloidal portions of the bed grade into each other through an intermediary zone with scanty development of amygdules. The amygdaloidal phase of the flow is usually covered by debris from the crumbling cliffs and drift material. The slope of the bed is always covered with grass–grown soil, through which little mounds of broken amygdaloid frequently stick up. In some of these the pieces show worn edges, while others contain pieces only with sharp edges and corners, indicating their connection with an amygdaloid bed at the place which is otherwise covered. Frequently these broken amygdaloid pieces contain small chips of native copper in the amygdules.

The matrix of the amygdaloid is dense and usually shows signs of alteration in various stages of progression from comparatively fresh to completely altered to epidote. In some places the alteration has proceeded to such an extent that only a crumbling mass remains with harder portions of epidote. Such is the case at Burnt Creek where an amygdaloid of this character was found, containing chips and flakes of native copper in the altered rock, and at Copper Creek where an amygdaloid outcrops, which shows prominently as a reddish mass with intermixed epidote. Some native copper was found in this bed also. The amygdules are filled with calcite, zeolites, epidote, chlorite quartz, a red mineral which probably is secondary orthoclase and native copper, one or more of these minerals filling the cavity. The amygdules show some variation in size and form. They are usually small, although amygdules measuring more than 6" were observed, in one place. At north shore of Dismal Lake an amygdaloid with elongated compressed amygdules, suggesting a viscous flow, occurs.

In places small fissures penetrate the beds, forming sometimes a network of small seams, traversing the shattered rock. They are filled principally with calcite, sometimes containing chalcocite. Fissures of this kind were observed striking nearly north and south and east–west. The conglomerate beds, which occur interstratified with the basalt beds, consist of pebbles of basic rock, pebbles with amygdaloidal development predominating. The matrix is apparently of the same material and is frequently permeated by calcite.

To the north of Copper Mountain a sandy shale overlies the basalt bed and is succeeded by fine to medium–grained sandstone, which continues north to the Bloody Falls, a distance of about thirty miles. Both shale and sandstone are of dark red to brown colour. The sandstone consists of grains of quartz and felspar, with a highly ferrugineous matrix. The felspar grains, which are smaller than the quartz grains, predominate. These sandstones are similar to the sand–stones in the "Nonesuch" group of Keweenaw series. The deposition of the sandstone was interrupted at four different times by eruption of basalt, which flowed over the floor and became interbedded in the sandstone. (Fig. 2.) None of these flows attained more than a few hundred feet thickness. The rock shows a somewhat coarser crystalline texture than the basalt at Copper Mountains, and the amygdaloidal phase of the flow is either scantily developed or non existent. Between the last two basalt ridges occurs thin strata of a greenish grey slate interbedded with the sandstone. Judging from the appearance the islands in Coronation Gulf beyond the mouth of Coppermine River and the capes on the west shore of the gulf are composed of basalt ridges.

## DYKES

At the foot of Sandstone Rapids a dyke, striking south 17° east, crosses the river. The dyke, which stands perpendicular, measures one hundred feet and consists of plagioclase felspar and a ferromagnesian mineral. It is possibly the source of the magma which formed the flows that are interbedded in the sandstone. The dyke shows cross–columnar jointing. The centre part is coarsely crystalline, gradually becoming finer in grain to glassy at the contact with the sandstone. The dyke–rock shows some copper–stain at the contact with the sandstone. The sandstone has been changed on both sides of the dyke from the backing of the hot magma. The red colour is changed to grey. The comparatively soft sandstone has become hard and fissile at the contact with the dyke. The change is visible, gradually diminishing to 150 feet from the contact. No data regarding the age of the Coppermine series were obtained. Petrographically they show great similarity to the Keweenaw series at Keweenaw Point.

### GLACIAL DEPOSIT AND GLACIATION

In the Copper Mountains evidence of ice action is everywhere present in the form of bed–rock scorings on the crests of the basalt ridges, which have been rounded off and polished. Occasionally a cliff shows scorings, indicating ice movement along the cliff. Glacial drift is to be seen all over the mountains. In the valley the lower benching and bed–rock topography is in some places concealed by small terminal moraines and till sheets, in no place reaching far up on the hillside. Most in evidence are the glacial deposits in the small basin around Tepee and Larrigan creeks and along the south shore of the Coppermine River on its north–west course through the mountains. The sandstone is covered by a thin sheet of till.

On the south side of the basalt ridge at Bloody Falls several hills of grayish white stratified clay lie against the basalt cliffs on both sides of the river. These hills, which are about two hundred feet high, have very steep sides, intersected by ravines, and present a very striking appearance.

### COPPER OCCURRENCES

It has long been known that copper in the native state existed in the northern part of Canada. It furnished the source of supply for the weapons and utensils of copper used by the Indians before they were supplied with iron from trading posts.

As early as 1771, Samuel Hearne on his journey to the Northern Sea established the fact that the Indians got their supply of copper from the Copper Mountains. He describes the "mine" he visited as situated about thirty miles south–south–east from the mouth of Coppermine River. That would be in the upper part of the series of basalt beds constituting the Copper Mountain, and would correspond to the eastward prolongation of the amygdaloid beds, which outcrop in the hill in the north side of Burnt Creek.

Later, in 1821, the Copper Mountains were visited by Sir John Franklin and Sir John Richardson. Their investigations were made on the north side of the big bend of Coppermine River around Stony, Glance, and Big creeks. They found many evidences of native copper, but as Richardson states they "did not observe the vein in its original repository, nor does it appear that the Indians have found it." Recently, Hanbury observed the occurrence of native copper in basalt on the islands in Coronation Gulf at Bathurst Inlet. These islands are probably the eastward continuation of the Copper Mountains. Even today the Coronation Gulf Esquimaux, or at least some of them, come to the Copper Mountains for their supply of copper. These people do not possess tools for working rock and are restricted to digging in loose rock for pieces of copper liberated through weathering and decomposition of the rock.

Judging from specimens in possession of the Esquimaux we met, the pieces usually found are comparatively small, although they occasionally find pieces large enough to beat out knives about 8" long and about 3" wide. In our search we did not find any large slabs of native copper. But in many places we observed small chips or flakes of native copper in the broken pieces of amygdaloid which forms small heaps in the flat soil–covered valleys on the back slope of the ridges. Usually more or less of a green copper–stain indicates the presence of native copper. In two locations or horizons we found native copper in place in amygdaloid beds, viz.: at the head of Copper Creek and in the hill on the north side of Burnt Creek.

At Copper Creek on the east side, well upon the hillside, an amygdaloid, showing an exposure of about thirty feet thickness, outcrops. The amygdaloid is much altered presenting a reddish appearance, which is noticeable at a distance. Kidneys or irregular masses of epidote occur in the bed. The altered rock shows a copper–stain on the outside, and although not abundant, small chips of native copper were found in this amygdaloid. It is, however, common that the broken rock shows copper–stained amygdules—(Specimen 145). The dense lower non–amygdaloidal portion contains tiny specks or shots of native copper. A short distance below this bed lies a conglomerate, but no copper was observed in it.

The most favourable locality for native copper, so far as our observation went, is at the north side of Burnt Creek.

A cross section of the hill is given in Fig. I. The bed marked No. I shows an exposure of about twenty–five feet thickness with frequent copper–stain in the amygdules—(Specimen 139). Red bands of a much altered rock with copper carbonate stain occur in the bed, which in places show cleavage like stratification. In bed No. 2 the rock, where exposed, has been very much altered in some places to epidote and a crumbling mass of light–coloured rock, in which nearly all the amygdules contain copper carbonate—(Specimen 140). Native copper in the form of chips and flakes is fairly abundant in this altered rock. In some instances a small un–oxidized chip of native copper can be observed enveloped in copper carbonate. Small fractures contain chalcocite. As far as the bed is exposed it shows a depth of about twenty–six feet. Above this amygdaloid lie three conglomerate beds. Of these the two marked No. 5 and No. 7 show a thickness of ten to fifteen feet and contain some native copper in the amygdaloid pebbles. The bed marked No. 6 shows an exposure indicating a depth of four feet. The two amygdaloid beds marked No 3 and No. 4, lying higher up, both show frequent copper stain, but no native copper was observed in them.

At Glance Creek, about a mile from its mouth, occurs what appears like a breccia but probably is the filling of a crack. It consists of altered basic rock cemented together with quartz, calcite and chalcocite. It outcrops irregularly in the bottom of the creek and on the east bank of the creek, where in places the adjoining rock looks like sandstone, stained with copper carbonate. No native copper was found here except as tiny shots in the hard basalt a few hundred paces to the east—(Specimen 149). Similarly at the mouth of Stony Creek the hard basalt shows native copper—(Specimen 148). Here the amygdaloid phase of the flow has been eroded away and is covered by drift where it dips under the mountain. Only in places the intermediary part, showing scant development of amygdules, remains.

POSSIBLE AREA OF COPPER BEARING ROCK

SKETCH MAP
SHOWING THE AUTHOR'S ROUTE
TO THE COPPERMINE RIVER

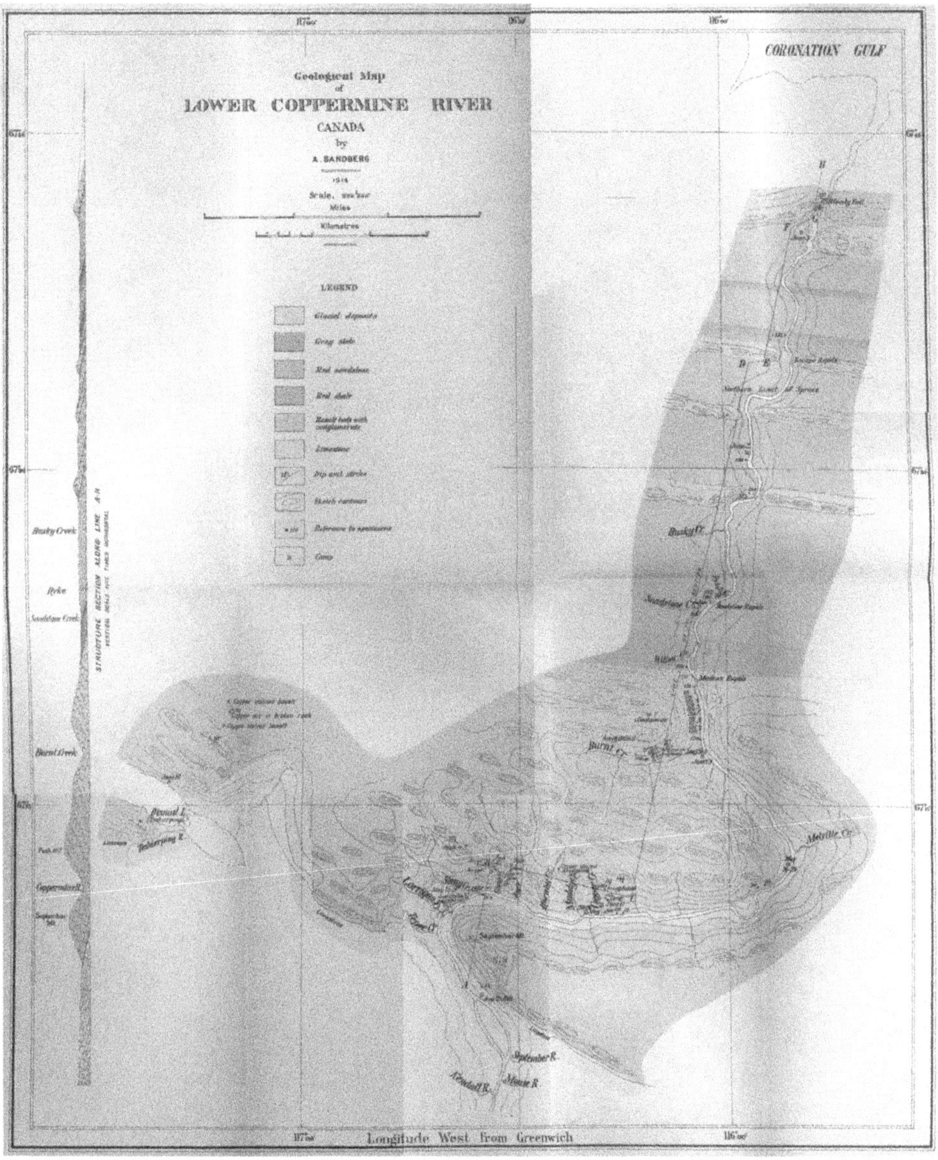

Geological Map
of
LOWER COPPERMINE RIVER
CANADA
by
A. SANDBERG
1914

Scale, 250000

Miles

Kilometres

LEGEND

Glacial deposits
Grey shale
Red sandstone
Red shale
Basalt beds with conglomerate
Limestone
Dip and strike
Sketch contours
Reference to specimens
Camp

CORONATION GULF

Longitude West from Greenwich

**Friday, July 28<sup>th</sup>.** Fine bright morning, we had breakfast at 8.30 then started final revision of suitcase and hand bag, getting out various small things for our trip. We took pictures of the canoe, finally made a start at 11.50. The river was a succession of small shallow rapids and quiet stretches, some of these 50-100 yards, and one long one of about 2 miles. I waded the Polaris up the rapids, the Dr. standing by to help when necessary. We had lunch at 1.40 at the end of a long quiet stretch; an Indian grave at the top of a hill, Saw the river ahead for a mile or so, and hills on both sides of it some seven or eight miles away. Saw some young geese, and when we started again I shot 4 of them with the little pistol. On again - frequent heavy work getting up rapids. A lovely afternoon, the river really very pretty and the timber much higher than I expected to see. We were using waders, but mine leaked badly, and chafed me a good deal. About 4.30 we passed the mouth of a little creek that I took to be the first branch to N marked on Bell's map. A little beyond this we came to a fork in the river, we were not certain whether this was an island or the N. branch, so left the canoe and went ahead to investigate. It proved to be quite an extensive island, perhaps a third of a mile, low and gravelly with spruce on it. We decided to go up the west channel, and waded the canoe up. The water was so shallow at a bar near the top of the island that it meant unloading the boat. It was then 6.10 p.m. so we decided to make camp. It was a very pretty camp, and we had the meal of the trip on stewed young goose. Very tired myself. Dried clothes, fixed up camp, etc. Went to bed about 10, but didn't sleep very well.

**Saturday, July 29<sup>th</sup>.** Fine bright morning, woke up at 5, but went to sleep again till 6.50. A comfortable breakfast, broke camp and got under way at 8.55. I discarded waders today, and tried overall pants only, but it was a doubtful change. We soon got into bad water, gravel bars and shallow rapids one succeeding another. At one of these we knocked a hole in the Polaris, we had to land, took out all our stuff, and made a patch of silk and brass. A most laborious day, and in the water the whole day, we did not paddle altogether half a mile. At some of these rapids we had to portage everything, at some to lighten the canoe of most of the load, frequently loading and unloading in the middle of the stream. At 4.45 we came to the box. We had had a ridge of hills terminating in a high bare hill, looked like sandstone and sand from where we saw it. The foothills to this were limestone, at the box the strata set right on end, one large detached piece that I supposed was Simpson's "Old Man of Hoy". We had to unload all our stuff at a bad rapid here and carried the canoe over. Loaded, hoping for a good stretch, but after passing a deep place with limestone set about 45° on the North side, we ran into worse trouble than ever, falls and bad rapids, impossible to either track or wade. We decided to make a portage of about quarter of a mile, and camped about half way, bringing all our stuff, including the canoe, to a very fine camping place. It was 7, by this time we were in shape to change our wet things and get something to eat. We were both quite weak from the severe labour of the day. Fried goose made us feel better, and I got enough energy to climb the hill behind our camp, hoping to see a caribou, but found to my surprise that we were camped only a short distance below the forks. We had estimated our days journey at less that five miles, and our journey of yesterday at less than eight, say 12 so far. Hanbury makes an estimate of the "Gorge" to Bear Lake as 10 miles, Bell's map makes it 20 miles to Hodgson's Point. Back to camp and put up the tent and turned in. My left knee troubling me where I fell today.

**Sunday, July 30th.** Awoke early and finally got up early - at 5.50. Breakfast and cleaned up our camp, and portaged our stuff above the falls, 220 yards from our camp. It took us just one hour, and we got underway, wading up several rapids, stopped awhile to investigate the South branch. Very little water flows into the main river here, a big gravelly bar. We were rather doubtful about it, but concluded to call it the South branch till we get some more data. Two or three more rapids brought us into a reach of quiet water. A very pretty river. We went up this for 20 minutes, the Polaris made good speed, I should say 4 miles an hour. We stopped at

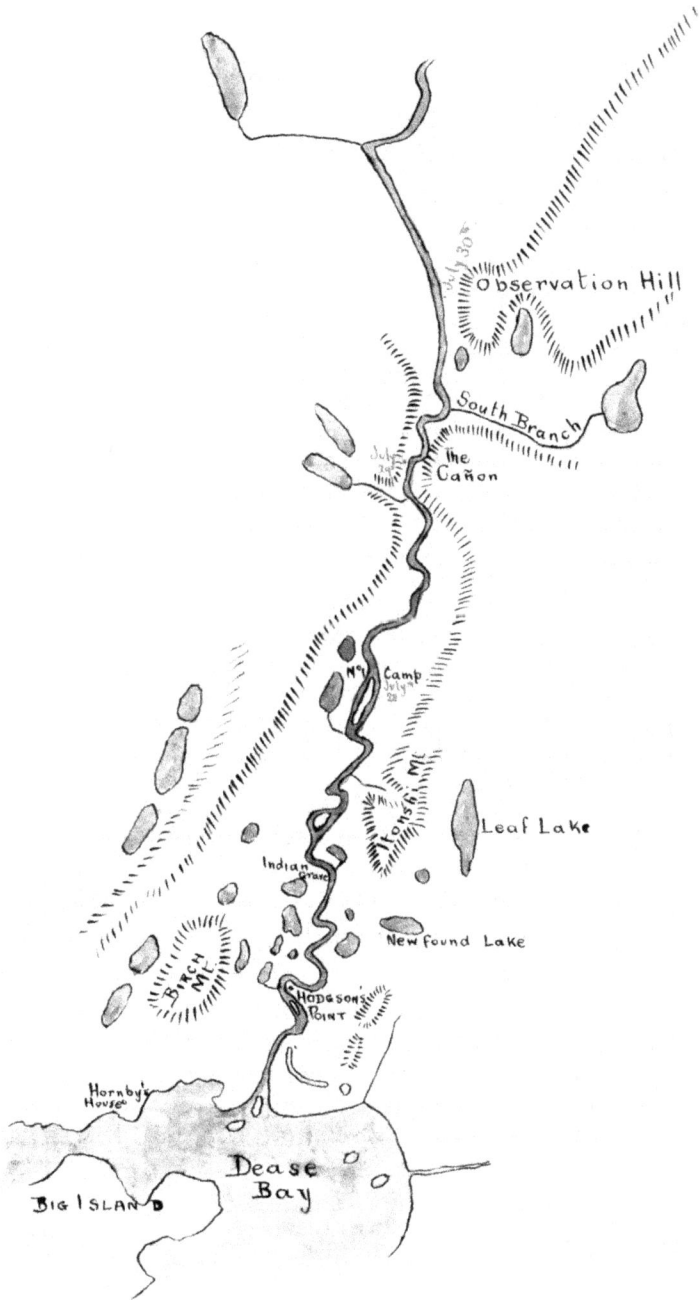

Observation Hill

July 30th

South Branch

The Cañon

July 30th

No 1 Camp
July 28

Leaf Lake

Indian Grave

Newfound Lake

BIRCH MTE

HODGSON'S POINT

Hornby's House

Dease Bay

BIG ISLAND

the nearest point to the large hill, and after some, discussion decided to make camp here, and investigate more fully our position. Put up tent and had a good lunch of fried hominy and bacon, then started out in full field order. A very warm day, I had on the lightest clothes in my equipment and used the skeleton coat. We left camp at 1.10 and got to the top of the hill at 1.40, stopping a few minutes to eat blueberries, lots of these on very small bushes, only an inch, or even half an inch, high. They looked as though they had been spilled out of a pail on the ground. Slightly acid to taste, no other flavour in particular, unless slightly of apples. Black flies numerous and troublesome all day, but a good breeze on the hill. We were both badly bitten, the Dr's face and my arms looked as though some one had taken a shot at us with a load of #12 shot.

Bearings taken from Observation Hill. All magnetic. Furthest visible point of Dease R. to S about opposite Bald Knob: S 31°36'W; Bald Knob: S 37°W. Big Island about centre: 25°W. Red Bluff above forks: S 42°W Mountain. N. shore Bear L: S 18°.30' (Subsequent obs showed this mountain to be on South side Dease Bay) Several small lakes in wooded country due West. A large lake between N 47° W and N 49° W. A small pond between above and river N 44°W. A straight stretch of the Dease R. still water no rapids visible N 40°W. and running, in a direction about 35°W. A large flat area, appearing from the hill like a basin, between N 45°W and N 4°E. Many small lakes in this. A red coloured hill, same character as Bald Knob being N. 38°W. Beyond this a prominent bluff.

We climbed ridge to South of Observation Hill. It proved higher, aneroid showing 550 ft. above river. Could see South shore Dease Bay, and Narrakazzae Islands quite plainly. What I had taken for a hill on N side was in reality some high peaks on the S. side. High mountains on N side not visible. Outermost Narrakazzae S 21°W. Small lake or large pond just over ridge, and another narrow pond end on at greater elevation. Walked over to next ridge again. Bar 630 above river. Saw a long lake running about east and west, we thought this might be Long or even Dismal lake (This "Long Lake" was some idea derived from Hanbury's book) the west end bears N 45°W, the east end N 26°W. From this hill we could plainly see the two big hills on the point near where we hung up last Sunday. They bear S. 40°W. Walked smartly back to our camp - it took just 55 minutes. The barometer came back to 20 feet of original reading. Tea, salt bacon and beans, then re-read Hanbury. Went to bed early.

**Monday, July 31.** Dull morning, slept very soundly. Dr. woke me as I was dreaming of L. G. and F. G. of all people [Later on, Douglas couldn't remember who these people were]. Got up at 5.40 and breakfast. Got off at 8.07. A dull morning with slight showers and sprinkling rain, soon turning to rain in earnest. The river was good traveling, banks regular, sandy, spruce much higher than I expected to find it. River course very erratic. Banks got sandier as we went up. In one place steep clay as well. Landed on a fine sandy beach at 10, thinking I had seen some geese. Saw some young ptarmigan, set out again, and came on a pack of geese that I had sure enough seen. We chased around after them on a shallow stretch of the river, and the first rapids or shoal water of today above. Finally got one young and one old one and then went ahead. A small stream, but larger than the so-called South Branch, came in due east at 11.30 (This was what we after called East River, or Steffanson R.) At 1 we came to an old Indian camp just below a rapid. The river had high sandy banks here, and we landed on a wide steep sandy beach, and investigated on the chance of there being a trail. Saw tracks of either moose or musk-ox [Douglas note, added later: It was caribou in fact, we did not at first realize how large a track the caribou leave]. No sign of any trail. Mosquitos and black flies both bad. We ate a lunch of cold boiled pork and bread, standing up in a shower, then waded up the rapid. We made good headway in the afternoon, though we had to wade up rapids occasionally. At three p.m. we came on a peculiar clay bank, and landed to have a look at it. From the top we could see three hills bearing N 40°W and Red Bluff N 30°W about. A little later on we came on a fairly recent camp on a stretch between two rapids. I supposed it to be some of Hodgson's, chopping of the same character as around Hodgson's Point. Apparently a winter camp. We went ahead, making good time though

wading more and more frequently as we went up. At 6 we came to an old
island, waded up rapids above it, and decided to camp there. A fresh NNE
wind and no flies. We made camp and had a goose supper. We were on
the site of an old camp here, perhaps one of Simpson's or Richardson's.
A lot of old wood ready cut. It came out a lovely evening, with a fine light
on the other side of the river, a pretty spot.

**Tuesday, August 1.** A fine, breezy morning, wind NNE true. The Dr.
got up at 5.40 and had the fire lit before I got up. Breakfast, old goose
stew, then got under way. Heel troubling me. We were camped just
below an island. River N 45°W compass. Fairly thick spruce on our
side of the river, more open on the other. Some detached hills bearing
N 40°W ahead. A short rapid, then about 3/4 mile still water, about
N 30°W then a long; shallow rapid N 80° W. A long shallow stretch,
we tried tracking, but it was too much trouble and risk, dodging
stones and shallows. Waded her up several rapids. At 10.30 we passed
a ledge of limestone, we identified this as the "Rock cut through by
river" of Hanbury. Another small rapid above it. General direction of
river N 40°W and without many bends, fairly straight course. More
gravelly shallows. At 11.15 passed an old camp, several teepees,
two lean-to's. A small quantity of short wood chopped, all very old,
probably one of Simpson's camps. More rapids above this, the banks
turning sandy again and large boulders above water in the shallows.
Came to a stretch of quiet water, strong head NNE wind. Landed
on a sandy island after a ptarmigan but didn't get it. Landed again
a little further up on a steep sandy bank, a clump of dead spruce,
country getting much more open. Had lunch, bacon and chocolate,
and on again at 1.55. Red Bluff still bears about N 40°W. On again
a fairly long stretch of quiet water then some shallow stretches we
had to wade up. At 3.30 a rapid where we had to portage all our stuff,
and more wading brought us into good water again! Landed at a point
with a spruce marked like a lobstick. Could see what we thought was
Hanbury's Kopje about N 40°W and 3 hills to NE. On again, the water
began to get shallow, and frequent wading necessary. River varied a

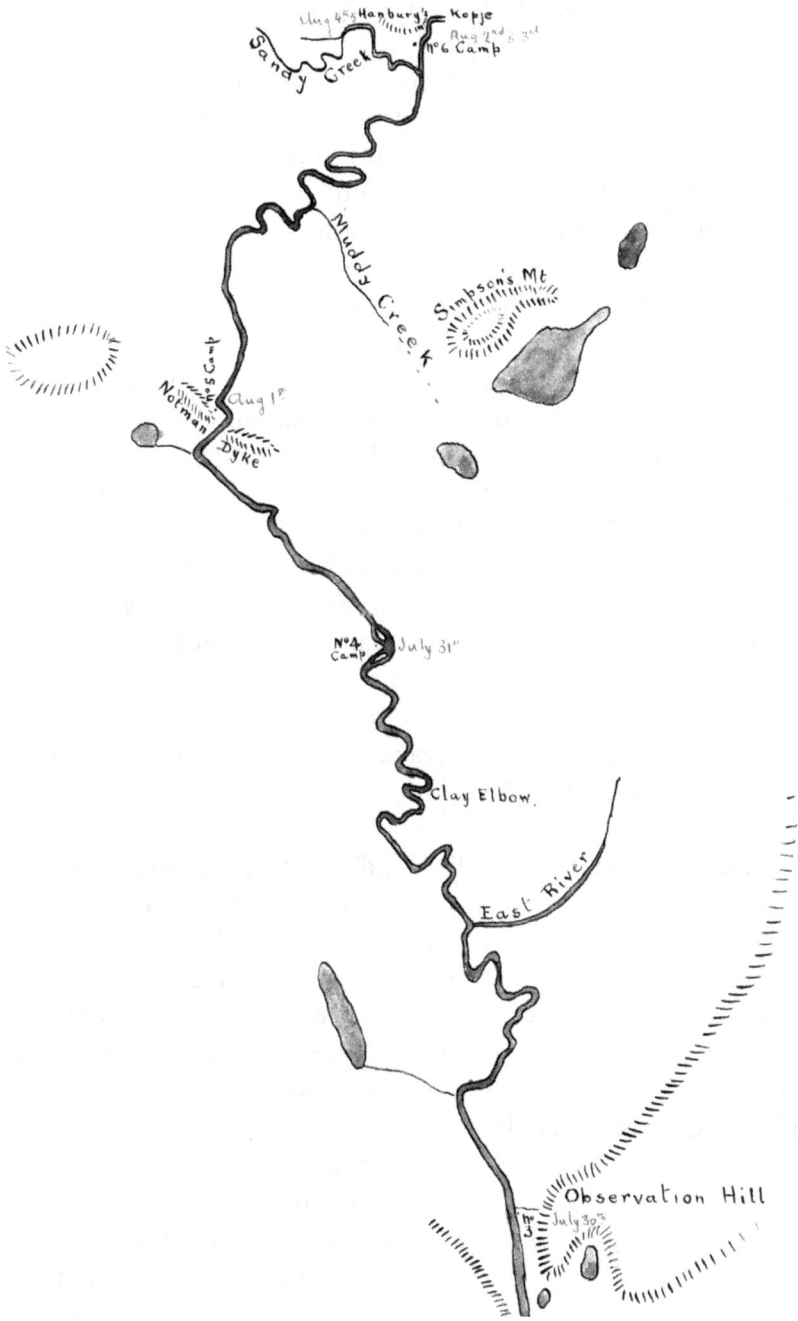

Sandy Creek

Aug 4th Hanbury's Kopje
Aug 2nd & 3rd
Nº 6 Camp

Muddy Creek.

Simbson's Mt

Forest Creek
Aug 1st
Norman
Dyke

Nº 4 Camp July 31st

Clay Elbow.

East River

Observation Hill
Nº 3 July 30th

great deal, but general course about N 10°E. At 5 we came to shallow spot, the river took an abrupt bend to the east and a small river came in from North. We landed at this point to investigate, the mountain that had been bearing about N 45°W was now to the west of us. The small stream drained a small lake about a mile from the river. Ahead of us was rolling dwarf-birch-covered country and another lake. To the NE and E was a long granite dyke with three high parts. The river proved bad here, a most exhausting wade for 3/4 of a mile, the Dr. going ahead with a pack. To my surprise the river swung around and cut through the dyke, no doubt that this is Hanbury's rock "cut through by the river". A sharp rapid above it, it was then 5.45 so we decided to camp, and moved all our stuff a short distance above the rapid. Cold, and strong NE wind, a bleak evening, and a feed of erbswurst [a compressed pea soup (peas, bacon, salt, spices) originally made for the Prussian army. It typically came 6 tablets to a sausage-shaped roll] was all right. We walked along the granite ridge but could see nothing definite about our position. To the N were quite rugged mountains, and a long granite hill, like the country to N of Lake Superior. Back to our camp and to bed early. We saw tracks of caribou, and what we took to be musk-ox, but no game in sight.

**Wednesday, August 2nd.** Dullish and strong NNE wind. turning fine. Temp. 46°. I woke at 6.15, we got breakfast and broke camp, starting at 8.45. Very bad, a succession of gravelly bars. At 10.30 we struck a sandy country again and dead water. We kept on, the river very erratic in its course, but no necessity for wading. A strong head wind. Sometimes the banks were high and sandy. It turned dull and cold, a strong NE wind — hard to paddle against. Stopped for a lunch of dry bread and chocolate by a high bank. Cold wind, could see nothing to give any idea of our position ahead, but behind, SW true, lay the ridge of granite we had camped by, distant I should judge 2 1/2 miles. After dinner we passed a high cut bank of sand, the river took a course very much to the east, sometimes flowing even S 60°W, 240° off our general course. The river became sandy, and at the same time swift and shallow, but the wind was as much with us as against us on the whole. At 3.15 p.m.. we stopped by a high bank to see if we could discover our position, and saw what was unmistakably Hanbury "Kopje-shaped hill" about three miles ahead. We tried tracking, but it was impossible. The banks kept high and sandy. Shortly after leaving the last landing we passed a stream coming in from the south, Quite a percentage of the total flow, and muddy water. The river becomes very clear after passing this, quite comparable with the Bear R. for colorlessness. We sometimes paddled, sometimes waded, and got to below the Kopje at 5.30. Quite a stream came in from north. The Doctor thought that this was "Sandy Creek". We kept on a little above it, and made camp at a very pretty spot, tall. spruce, all around. It came on to rain, we were cold and wet, but made tea. I had shot a couple of young geese and a small duck, we had these and apples, quite the feast of the trip. It rained very hard, woke up frequently during the night.

**Thursday, August 3ʳᵈ.** Dull and stormy with frequent showers. Heavy low clouds and NNE wind strong. Temp. at 9:45°. Got up 6.30 and breakfast, everything wet and sandy. We made preparations for starting out and finding where we are at. We went up the river first and climbed the kopje from the point 70 feet high. We could see very little definite from the west side, some stony rapids on the river about 1/2 mile up, and numbers of lakes, From the NW side we could see even less, there was much rain and driving mist. We went back to the canoe and paddled up to the first rapid, then walked about 2 1/2 miles along the river. It was of the same character throughout that distance, short shallow rapids over gravel. The country has got quite devoid of trees, sandy, and in places granite. Back to canoe, shot a couple of Arctic trout, then to camp in rain. A hurried lunch, then started off to explore the north branch. This was small, we could hardly get up in the empty canoe. We landed on south side near mouth on the chance of seeing a trace of Hanbury's camp. We came across an old camp, trees chopped down, four trees trimmed and a notch cut in them about 5 1/2 feet up. This did not look like Indian work to me, but it was very old, at least 20 years (It was probably some of Simpson's). We could not go very far without wading, so we left the canoe and walked along the bank. The river was sluggish and sandy. Presently we came to where it divided, this was unforeseen and perplexing. We decided to strike for the high hill to NE (compass) on the chances of being able to see Dismal Lake. We had a hard walk of about 4 1/2 miles over rolling country, lakes all around, spruce here and there, some tracks of what might have been musk–ox (It was caribou) but though we commanded an extensive tract of country nothing living to be seen. Walking wet, sometimes swampy, sometimes thick dwarf birch, which was very wet. It rained all the time and was quite cold, 42°. We got to the granite mountain and found another ridge ahead, climbed this and found it ran for some miles NE. No sign of Dismal Lake. This ridge forms the division between the two branches of the Dease R. It was an unspeakably gloomy view, lakes everywhere and heavy black clouds—mountains to the NW hidden by clouds. Both sides looked bad. it was a hard outlook. I made the hill 470 ft. high by barometer. Back again. lots of blueberries. Got to canoe and to camp at 6.50. Made a good fire. had our supper of trout, and dried out our things as well as we could. Looks as though it would clear tomorrow.

**Friday, August 4ᵗʰ.** Awake at 5.20 and up at once. Got everything fixed up and broke camp at 8.30 a.m. Temp. 7.30: 39°; at 5.30: 44°. It looks very threatening, heavy clouds coming over hills, strong NE wind and slight rain. We went down to the fork and started up what we took to be "Sandy Creek". The water was shoal, all sand, and frequent quicksand bars. But we got along fairly well, several good paddling stretches, and we had three heavy bags arranged so that we could lighten load quickly. We had to do this several times, then came to a long stretch of good water with high sand bank to NE side. At this point the river branched in two, they looked about the same size. It was small, the paddling was fairly good. It could not possibly be Hanbury's "Sandy Creek". We decided that the only thing for us to do was to go ahead on foot, and find out where the Dismal takes lay. So we came back to the junction of the two small streams and made a comfortable camp as we may be here a day or more. We estimate the distance to Dismal to be 15 miles. We fixed up camp and dried out our wet things. The weather cleared up and it came out a fine afternoon.

I did some overhauling of our gear, washed some clothes, washed out the Polaris and then at 3.15 P.M. started for Hanbury's Kopje for another look around. It took me just 35 minutes to the junction with a strong favorable wind, and not having to wade at all. Climbed up the hill but could see no more than I could yesterday. I searched the plains thoroughly with my glasses, an extensive view and a fine clear day, but not a sign of deer. Bearings from Hanbury Kopje: saddle of Granite Ridge: N 29°E; western spur of Granite Ridge: N 26°E, eastern spur of Granite Ridge: N 42°E. A large bluff, the farthest and most prominent hill visible from here something like this:

Point "a" bears N 52°E. The two spots of snow on ridge to S ....S 51°E. As far as can be seen the general course of river runs towards the North end of this long ridge, or N 78°E. Back to camp, quite a struggle with empty canoe and shallow water against a strong head wind. We had a comfortable and leisurely tea and to bed early. A lovely sunset, but it looks like rain tomorrow.

**Saturday, August 5th.** Awake at 5.20 and up and breakfast. A dull day threatening rain. We started out nevertheless at 6.53 to try for Dismal Lake. It came on to rain a few minutes after we started, very cold and a few flakes of snow, thick clouds and mist from NE, the direction of the wind. We got as far as the granite mountain about 8, nothing but mist in every direction, we decided to wait until 12 on the chance of it clearing off. It was rainy and blowing, very cold and raw. While waiting under the lee of a big rook a big bull caribou came along. The Dr. shot once high, I also shot twice high, and shot him through the jaw and neck the third shot, about 175 yds. away. He was much larger than I expected, must have weighed over 250 lbs. (Probably nearer 350) We butchered him, an awful mess. At 11.30 it showed some signs of clearing so we decided to walk along the top of the ridge. It came out finer, still there were frequent showers, but we were able to get a view of the country. We could see nothing to indicate the lay of Dismal Lake, and it was too far to go on to the next ridge from which they might have been seen. In any case a fairly large lake intervened. We decide to take the right hand stream, it can be seen to come out from a lake, this is probably Simpson's route. A most intricate country, lakes and sand hills in every direction. We walked back along the ridge in a heavy rain, and more or less rain till we got back to our caribou. We arranged a pack of meat the hindquarters etc. swung on a pole, but we had to give that up as it was too heavy. So we cut off some of the meat and packed it into camp - a laborious job, though I had only 31 and the Dr. 37 lbs. The caribou must have weighed nearly 300 lbs. We had a caribou supper, a rather fine evening and to bed very tired and not very cheerful over the prospect of the next seven days. From end of granite ridge, Aug. 5th, river general course east (true) ends in a long lake with a ravine on the east side. Spots of snow bear S 38°E.

**Sunday, August 6th.** Dull morning, raining at 2.30 and still raining when we woke up at 5.40. Raining at breakfast. We packed up our stuff, tying some of the smaller things together for ease at the portage and made a start at 8.45. Got to the junction at 9.30, some heavy pulls, but did not lighten the canoe. Passed our old camp at 9.45 then dull, NNE wind and raining a little. (note added later by Douglas: We stopped to make a cache just above our old camp on the opposite side, but no mention made of this in journal). Got into gravelly country, frequent wading and lightening of canoe necessary. Kept at it till 1 when we had lunch. At 2.05 we passed the spot as far as I had gone on this stream the other day (Note added later by Douglas: Refers to the walk the Dr. and I had made on Aug. 3rd. The following November we found the footprints we made on this occasion as sharp as the day on which they were made.) Conditions improved thereafter, sandy banks and flat gravelly bottom, the characteristic of this branch of the river. At 3.50 we passed a small stream coming in from the east (N 20°E). I landed, it seemed to drain a large flat with several lakes. Above this we entered a lake at 4.30 p.m. The river left at the same end it entered. This lake was surrounded by

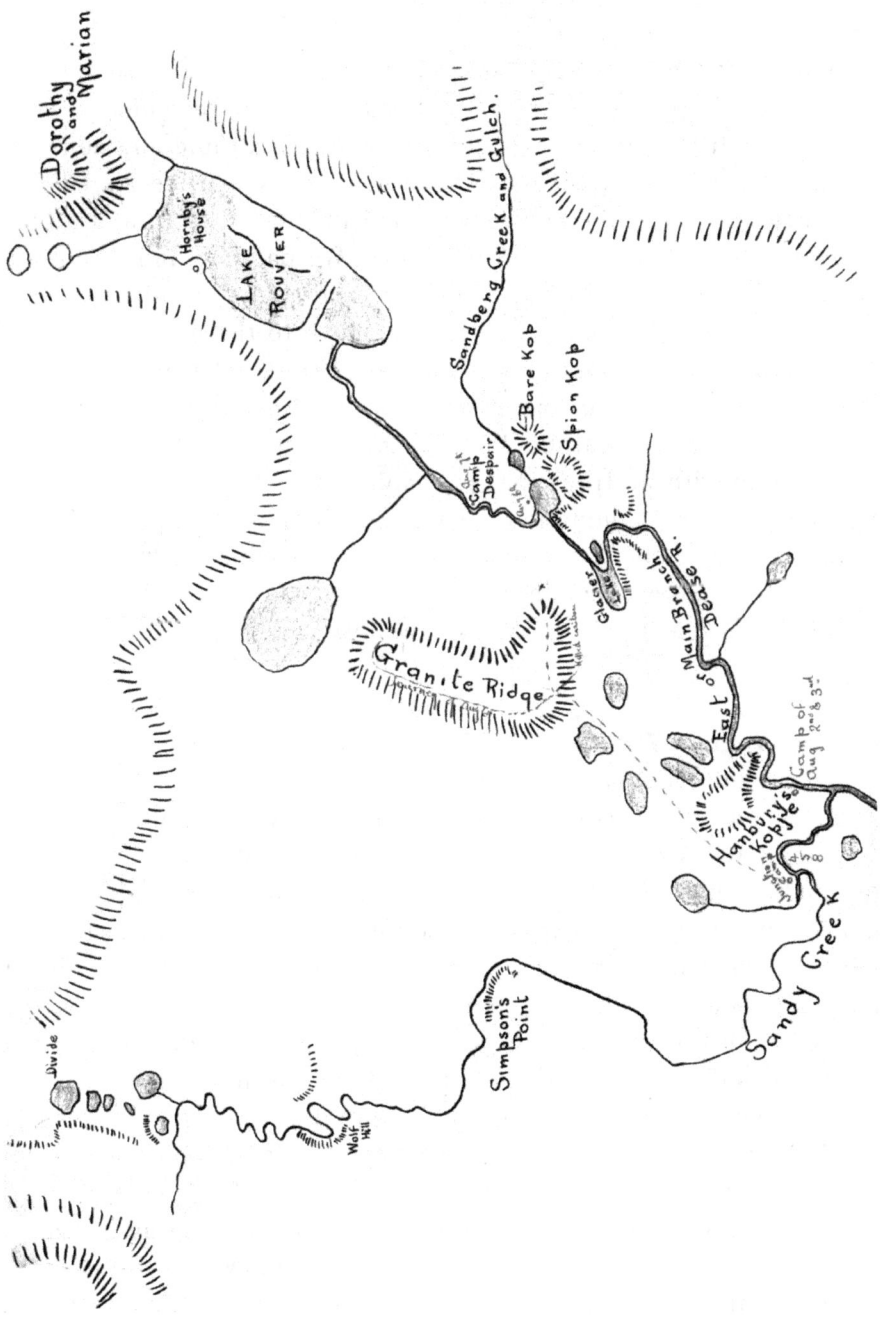

Dorothy and Marian

Hornby's House

LAKE ROUVIER

Sandberg Creek and Gulch.

Bare Kop

Spion Kop

Camp Despair

Glacier Lake

East to Main Branch Dease R.

Granite Ridge

Hanbury's Kopjes

Camp of Aug 2nd & 3rd

Sandy Creek

Simpson's Point

Divide

Wolf Hill

rounded sandy and gravel hills some 50 or 60 ft. high, the one at the
entrance 35 ft high. The river got very shallow and spread out over a wide
bed but very little current. At 5.10 we waded up a rapid and considered
camping at a good spot above. There were more spruce ahead, so we
decided to go on to this. A high rounded hill to NE of it, and a turn of
the river brought the Kopje #2 into view. We paddled and waded on to
this, there is quite a bay with a little lake and a small rivulet entering it.
It was a mere trickle, the river went on more to the north. We camped
at 5.55 opposite the Kopje #2. This was a rounded gravelly hill, well
covered on N side by spruce. A wet cold evening. High NNE wind and
showers, detestable weather. Walked up the more northern of the two
hills, the one with no trees. NO probability that the other is Hanbury's
Kopje on further examination. From the summit: Hanbury's Kopje: S
64°W. and distant about 7 or 8 miles. The river course appears to trend
about N 48°E. The lake cannot be seen, a rather high gravel ridge, glacial
moraine, intervenes. The saddle (Sandberg Gulch): N 79°E; the snow
spots: S 47°E

**Monday, August 7th.** Awoke at 5.30 to find it raining hard. It let up a
little as we got breakfast, but was raining when we broke camp. I had a
regular catastrophe this morning in getting one of my boots burned, the
only pair I have. We made a start at 8.30, and went up some very difficult
river, making not more than half a mile in a straight line till 11. We had
to take everything out of the boat a couple of times and most of it all the
time. At 11 we came to a bend, I went ahead, the prospect was appalling,
the river spreads out over a gravel bed making it straight portage. The
Dr. and I each took a pack and went on up the river. We went up about
one mile with the packs, the river was quite impossible. We left the packs
and went over to find first a high ridge hoping to see some sign of the
lake. Another similar ridge of gravel moraine lay ahead about a mile.
We walked on to this and could see one corner of the lake about five
miles ahead. The river was of the same character right along, impossible
to take a canoe empty even. This is a severe blow. We discussed the
situation in all its bearings, meantime walked back to our packs in rain

and N wind. We took them back to the canoe, the situation is extremely puzzling. This cannot be Hanbury's "Sandy Creek", it may be Simpson's route. I should say it leads to the large lake marked on the Admiralty chart towards the U of the Dismal lakes. We cannot possibly go on this branch of the river. It will take at least 5 or 6 hard days to get to the lake, and even then we do not know whether there is any short pass over the divide. The mountains behind which the Dismal Lakes lie are still 15 miles distant and of a rough character. We came back to our canoe, and made camp in a pouring rain and strong N wind. We could put the tent up only on the point of the gravel hill, exposed to all the wind there was. We talked the question over, and finally came back to the idea of trying the other route. It is the best we can possibly do. A late lunch, then fixed our camp. Overhauled our gear and had a game of crib, but we were both tired, and I was too worried to care about the game. Temp 38°. We had an early and sumptuous tea, fried caribou, bread, and rice with raisins, the best of our stores, and turned in early.

**Tuesday, August 8th.** Spent a worried and restless night, waking frequently. Went out at 2.30, it was nearly sunrise and a cold morning with low-lying clouds and a streak of blue sky to the NE and strong NE wind, very cold. Tent frozen—temp at 5: 34°. We broke camp, tried pack harness etc., and finally made a start with loaded canoe below first portage at 8.10. Passed our old camp at 8.46. A few patches of blue sky and many low fleecy clouds, persistent NNE wind and cold. The Dr. walked along the shore while I took the canoe down shallow sandy stretch of river. We passed through the little lake at 9.50. It runs N 65°E; S 65°W compass. Passed the little creek at 10.07 it runs due E and W compass. Between there and the small branch from SE the limit of our walk, there is mostly good water, although we had to lighten once and wade a good many times. We got to the small branch at 11.05, and lunch at our same old place, boiled caribou meat and Dease R. sandwich (This was a thin slice of cold boiled salt pork between two slices of boiled caribou). The sun came out at lunch, it felt very pleasant, and at once mosquitos began to show up. We went down the next three miles in pretty good time, Got back to the fork at about 2.30 and up to our old camp. Shot some fish on our way up. We made a comfortable camp, and overhauled our stuff, caching what we could. Got to bed early.

**Wednesday, August 9th.** Fine bright morning and warm, the first fine morning since Sunday a week ago. We got up at 6, broke camp, and were under way at 8.30. We took the left hand of the two branches, and had to wade and pull the boat most of the time. At 10.00 we got to stony rapids, impossible to take the canoe further. I should say we were by water 4 miles from the junction, or 2 miles in a straight line. I went ahead a little, nothing for it but a long portage. The country was decidedly pretty on a fine day like this, a gently undulating expanse, grass and dwarf-birch-covered, good to walk over for the most part. I walked ahead about 3/4 of a mile, the creek seemed the same, rounded stones. We made a start at 11.05 and made the following: 1rst lap: 500 yards, 2nd: 300, 3rd: 350, 4th: 300, 5th: 200, 6th: 320, 7th: 300, 8th: 420, 9th: 300, 10th: 350, 11th: 300. Total: 3640. We stopped for lunch at 12.45, the Dr. was unwell and couldn't eat anything but said it didn't interfere with his packing. It was fearfully hard work, about the 9th lap I felt pretty near all in. We used to make 4 trips, 3 heavy and 1 fairly light. The weather was lovely, the flies were something fierce, both black flies and mosquitos. We carried on to a point with a few small spruce, then decided to go on to a good camping place about 300 yds further on, a clump of spruce on a point, and a still reach of the river above it. Evidently an old camp, I should judge of Simpson's. We made a sumptuous tea, bread, caribou steak, apples and rice, and we sure needed it. Just as we finished I saw a bull caribou on the other side about 1000 yds away. I ran back along the path hoping to head him off, but he had taken a notion to hike for the granite ridge. Back to camp very weary, cleaned up and made our beds, and turned in about 10.30. The evenings are closing in now, at that time it gets difficult to write in the tent, but no stars visible then as yet. Bearings from our camp on the sandy hill: Sandy Creek (Simpson's Point) Aug 9th, Hanbury's Kopje: S 42°E; snow spots: S 64°E. I was not sure of the course of the river, there appear to be two possibilities ahead. A high sandy bluff with spruce woods bears N 54°W.

Possible divide A: N 62°W; B: N 20°W. The sandy bluff is distant about 2 1/2 or 3 miles, the river apparently runs close to it at least. Northerly end of granite ridge: N 60°E; southerly end of granite ridge: S 80°E.

**Thursday, August 10th.** Fine quiet morning, slight haze in sky. Got up at 6.30 after a troubled night, dreams of l.g. tanks etc. Left for canoe at 8.20 distant by pedometer 2 miles 280 yds. By pacing 2 miles 200 yds. Very hot, I was wearing light waders. Flies very troublesome, both black flies and mosquitos. A heavy and a hot job bringing up the canoe, but we had a few paddling stretches. We got back to our camp at 10, packed up our stuff and made a start at 11. The river was deep and sandy for a while, the sky became overcast and a fresh south wind. We got to a bar of rock where it was necessary to make a complete portage at 11.20. Put the stuff in again for a few minutes, then came to another rocky stretch. We made a complete portage of the stuff and took the canoe up the river. Portage 450 yds. Then another short complete portage, and we loaded the canoe and had lunch. Made a start again at 1.20 p.m., the river was deeper. Strong S wind and turning colder. The good water did not last long, it got worse and worse, heavy pulling all the time. no paddling, frequent lightening of the canoe. By 5 P.M. I was

nearly all in with the heavy strain. We were then near a high spruce covered bank with a bare sandy hill covered with dead spruce trees to our right. We left the canoe and walked ahead, the river makes some great loops here. We decided to camp, so portaged most of the stuff across and carried it up, and waded the canoe up. A pleasant place for a camp, and a lovely evening. We had erbswurst. I felt very tired, but took a walk after we had fixed up camp. Skirted along the top of the hill, saw no deer, nor deer tracks, but many wolf tracks, and saw a little wolf. It was kind of sandy amphitheatre. The view up the river was very lovely, the high sandy bluff then a valley of small sandy hills with a good deal of spruce. Back to camp, fine sunset and clear western sky.

**Friday, August 11th.** Fine morning, turning dull and cloudy and strong east wind. A few drops of rain now and then. Got up at 6 feeling very sore all over. Breakfast and broke camp, made a long portage across two of the bends, then brought the canoe up. We couldn't take any load in river, so made another long portage of the stuff across to a high sand bank, and I took the canoe with some of the stuff around. The creek looked better above. We climbed up this high bank of sand to get a picture, and to look around. It was still hard to trace this winding ditch, but it can't go much further now. Divide, or what appears to be it bears N 11°W, Hanbury's Kopje: S 44°E. Started again at 10.59. We got on fairly well till 11.30, the river was very narrow and made many bends, but it was deep. Quite a job to get the canoe around some of the corners. About 11.30 we came to a little rapid, we landed to look around. The Doctor took to the right and I took the left hand and climbed a small hill. From here could see a small lake apparently near the divide. This seemed the best place to investigate from, so we had lunch at the canoe, then pushed on. One complete portage, and several lightenings of the canoe brought us to a good camping spot near the lake. We fixed up camp at once, and made a start to look for Dismal Lake. Dull and cloudy, strong east wind. Bar. 28.81, T. 65°. Made a start at 3.10, course N 10°W. comp. Got to a saddle along a ridge at 4.10, 2 miles, 1300 yds. Sighted water ahead bearing N 10°W. Elevation above our camp: 180 feet. Dr. walked along the ridge to NE while I took the valley. Undoubtedly Hanbury's "Ravine". We could see the lake quite plainly, but more distant than Hanbury's estimate. We walked along to the end of this ridge and had just decided to go on to the lake and for Sandberg to wait for me when I happened to see a man at a camp. We walked across to it distant 1 1/2 miles from the ridge and found him to be an eskimo. He did not see us until we

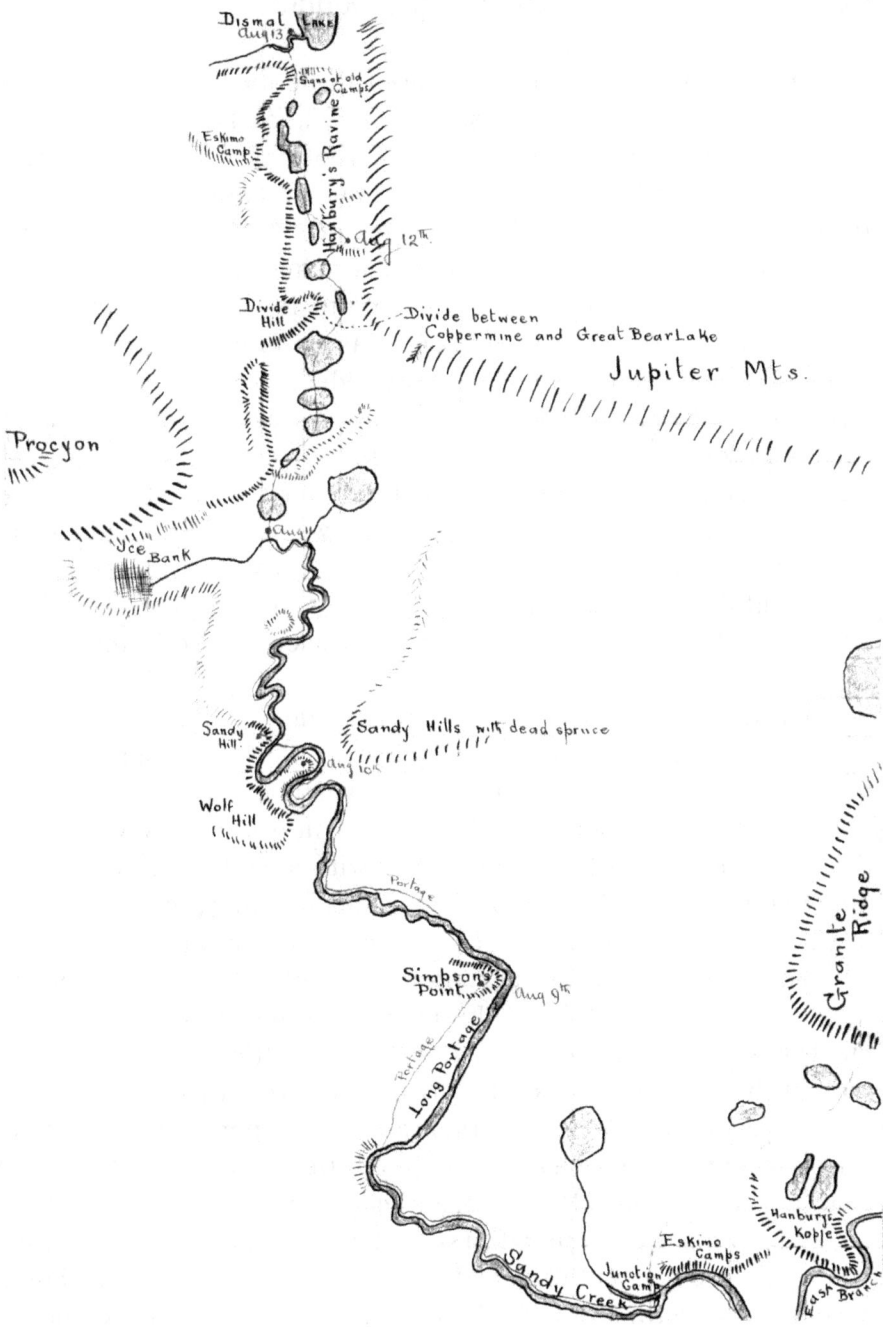

Dismal LAKE
Aug 13

Signs of old Camp

Eskimo Camp

Hanbury's Ravine

Aug 12th

Divide Hill

Divide between Coppermine and Great Bear Lake

Jupiter Mts.

Procyon

Ice Bank

Aug 11

Sandy Hills with dead spruce

Sandy Hill

Aug 10th

Wolf Hill

Portage

Simpson's Point

Aug 9th

Portage

Long Portage

Granite Ridge

Hanbury's Kopje

Eskimo Camps

Sandy Creek

Junction Camp

East Branch

were quite close and at first appeared to be much frightened. We held up our hands and he did the same, then shook hands and he appeared to be quite reassured. We could make nothing out of him except that this was Teshierpi Lake. We made various attempts and signs to get him to come back with us. and help us pack our stuff and he did walk along a short distance, then turned back. I watched him through my glasses make back for his little camp, and off to the NW with his outfit. He was a short muscular man, rather pleasing face and clean, dressed in short skin pants, short skin coat, fur inside, and sealskin boots. He had a switch in his hand, and I saw a bow and some arrows in a case, and some spears under a roll of fur supported on four crossed sticks. We walked back to our camp, stopping to investigate for a route at this side of the divide. There are several small lakes that we can utilize. Back to camp at 8.35, rather tired. We had walked just 16 miles. I should say the divide was about 6 miles in a straight line. Tea, potatoes, and coffee. To bed at 10.45. Saw the moon also two stars, and used candle in tent at 11 p.m.

**Saturday, August 12<sup>th</sup>.** Fine bright morning, up at 5.50 and breakfast: hominy, bacon, potatoes, and new bread, We broke camp leaving the things ready for portage. and went to investigate the ice seen yesterday. The Dease R., or Sandy Creek, runs up into a narrow valley, S 80°W and finally ends in a large gravel flat, with a bank of ice on north side, a miniature glacier. Spruce on both sides of valley, lots of mosquitoes, fresh east wind, and hazy to east. Back to our camp and made a start from there at 9.10 a.m. We portaged to the little lake, then over the hill, very heavy work, and down the long grade on the other side. We had lunch beside one of the lakes. I shot three little ducks. On again p.m. around the "big" lake near the divide, over the divide, and into a lake on the Teshierpi side. It was then about 5.10 p.m., so as we had done very well, we brought our stuff up to a little mesa about 40 yds from the boat and made camp. Heavy clouds came up and the wind, which had been SE all day came around first to the west, then to the NE. We had a shower before tea, and it turned very cold again. Before the shower the mosquitos were thick, I had a hell of a job to clean the ducks on account of them. Took a stroll over the gravel mounds behind our camp, but saw no deer. Stormy looking evening, to bed early and a good sleep.

**Sunday, August 13th.** Dull, raw, and cold, heavy clouds, the mountains across the ravine, and on the other side of Teshierpi Lake covered with mist. Strong NE wind. Up at 6, breakfast: potatoes and bacon, then broke camp. Made a trip with one load about 300 yds, then back and got canoe along so far 660 yds. Made another lap of 550 yds, finished this at 9.50. Then 405, and 360. This brought us to the triangular lake, then we portaged into a longish very shallow one, then into the largest. We had lunch on this one, on the east [map shows west side] side, at 12.15, and after lunch walked to the top of the hill. I was much dismayed to find that we were still a long way from the lake, it was uncertain which side of the hill was the better, so we went to the end and decided to take the west side. We made it 1000 yds to the crest, and 600 more to the canoe, so started in, both of us much fatigued by the heavy work of the morning. The lake helped us for about 330 yds, then came straight carrying for about 13 or 1400. We got over this somehow, and took the canoe down to the lake, or rather into a little creek running into the lake, at 4.30 p.m. The last stages showed signs of extensive camping, all very old. I thought it likely to be some of Simpson's. We pitched on a camping place not far from the lake about 100 yds, and close to the little creek at the mouth, then made the last wary lap, got our tent set up and changed, and had a sumptuous supper in honour of the event, <u>Raspberries</u>, bacon and potatoes, new bread, and rice, the latter slightly burned. We took things easy after—a great relief to have this portage over. A dull stormy evening, strong NE wind and rain.

**Monday, August 14th.** Dull rainy morning, NNE wind. We did not get up till 7.30, when we broke camp leisurely and made a start at 10.25. Course

N 20°W compass. Passed point of island 10.47, changed course to N. Off Sandy Point at 11.07, changed course to N 12°E Point where we landed at 12, then course N 30°E. Lake averaged about a mile wide here. Next 12.12 changed to N 41°E. This was rather a steep point say 30 ft. high. At 12.20 changed to N 50 E. Landed at 1.40 for lunch, on at 2.10 N 60°E. Passed Dismal Point. Watch stopped for 30 minutes. At 3.50 entered a narrower and long part of the lake bearing due east. We paddled down this long stretch in drizzling rain and NE wind, and got to a high sand bank at about 6.15. A high sandy bank on S side of lake and a shallow

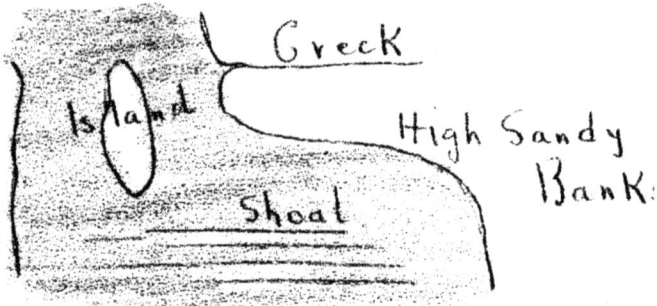

sandy bar stretching across. We had to wade, and landed on the islands a low willow-grown sandy bar. The beach was strewn with the bones of deer. It looked to me much like the end of the lake as described by Richardson. particularly as a small creek came in below the sandy bar. This was a small creek, about the same flow as Sandy Creek. We had to wade the canoe in several places. The water had the character of a river rather than a lake, the shores were rather high, rounded, willows in places near the water, and grass. At one place we came across a double line of stones with tufts of grass on top of them, quite recent. I took them to be Eskimos work, with reference to hunting deer. We decided to camp, so tried first the N shore, but it was too marshy, so to the south side and pitched our tent on a very comfortable spot. It was raining hard, we turned in at once, eating a supper of hardtack in our bags. I found it cold, my sleeping bag was wet, and the tent was leaking, but dropped off to sleep.

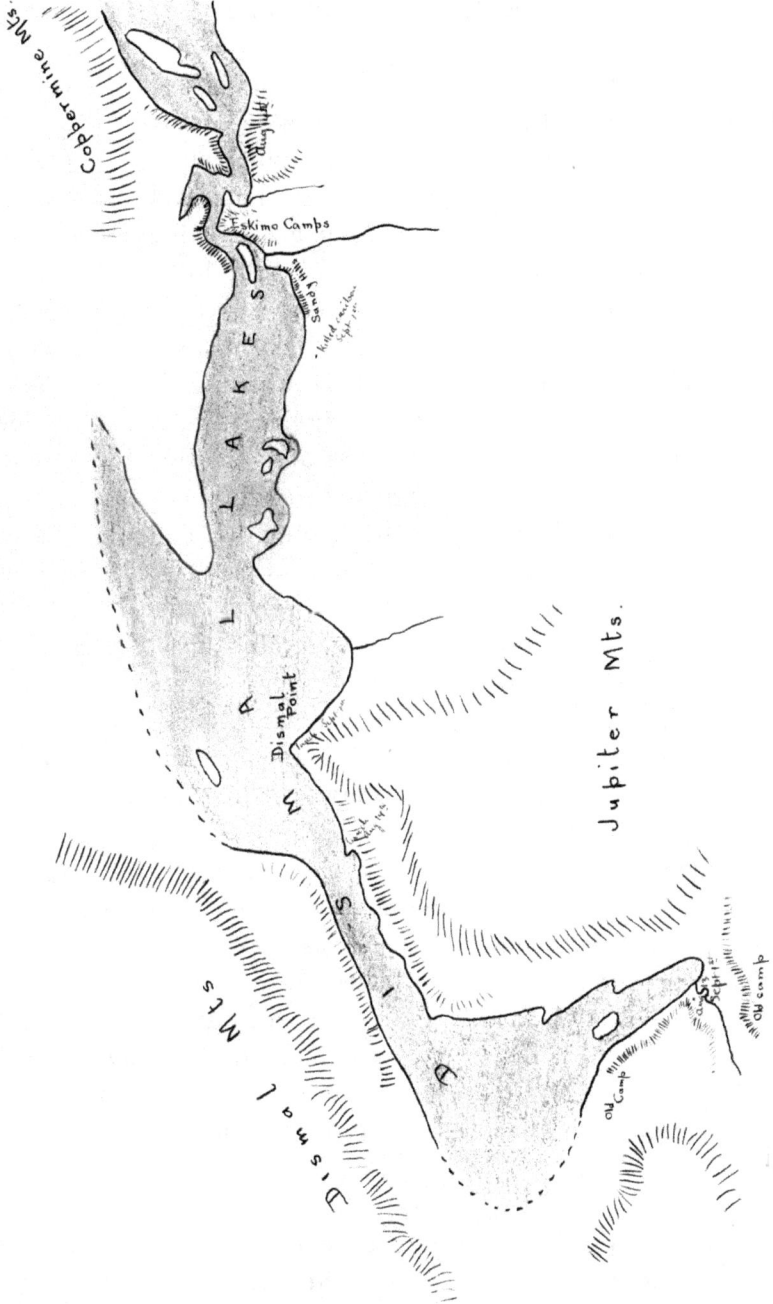

Coppermine Mts.

Eskimo Camps

Sandy Hill

killed reindeer
Sept 1910

L A K E S

Dismal
Point

A

M

Jupiter Mts.

Dismal Mts

S

I

G

Old camp
Sept 1

Old
camp

**Tuesday, August 15ᵗʰ.** A lovely fine morning with a few low-lying misty clouds. I was awake at 2.30, but went to sleep again till 7.30, when we got up and made breakfast of erbswurst, our first experience of boiling a pot with willows, but we got on all right. This part of the lake looked very pretty this morning, the rounded grassy hills brilliant after the rock and mist of the last few days. Climbed the hill and noted the lay of the narrows, about 1 1/2 miles or so long. Saw an eskimo camp to the west of us. We packed up our things, there came on a heavy shower, then we went to look at the camp. There was a kayak with stones piled around it, a circle of flat stones on edge, a stick with the sleeve of a jacket hanging to it, and a little further on a paddle and a stick. We left a file and a needle and back to our boat, made a start at 12.30. A fresh west (compass) wind. Till 1.10 N 60°E then N 75°E. We landed at 1.25 for lunch and got under way at 2.00 p.m., course same. High hill with rocky top on north side of lake, and also high, but gently rounded, grass covered land to south. Some high mountains bear N 70°E. Kept this course till 2.27, when we opened out another long stretch of lake, a very pretty view, beautiful sky and fresh SW wind, course N 87°E. (Course should be S 75°E bearing of Kendall R. from point). Got to this point at 3.30, it proved to be a channel between this and another lake very shallow and sandy. Pushed through wading, and paddled down the next stretch, arriving at end at 5.05. Course S 85°E. Stopped to investigate the Kendall creek, it was a flow about as big as the Dease above Hanbury's Kopje and comes in from S 60°E compass. We went a short distance down the river, and camped below a little rapid. There is not much water in this river, and rocks are dangerous. I got out the net while the Dr. made camp, a hell of job, but I stretched it out somehow, and then took a troll. I hooked some kind of a fish which I lost. Rice and apples only for tea and went to bed early. A rainy night, and much depressed over the outlook generally.

**Wednesday, August 16ᵗʰ.** It rained hard all night and was raining at 4.30 when Sandberg woke me. I looked at the net and found one large fish, whitefish?, then took my rifle and went for a long hunt on the other side

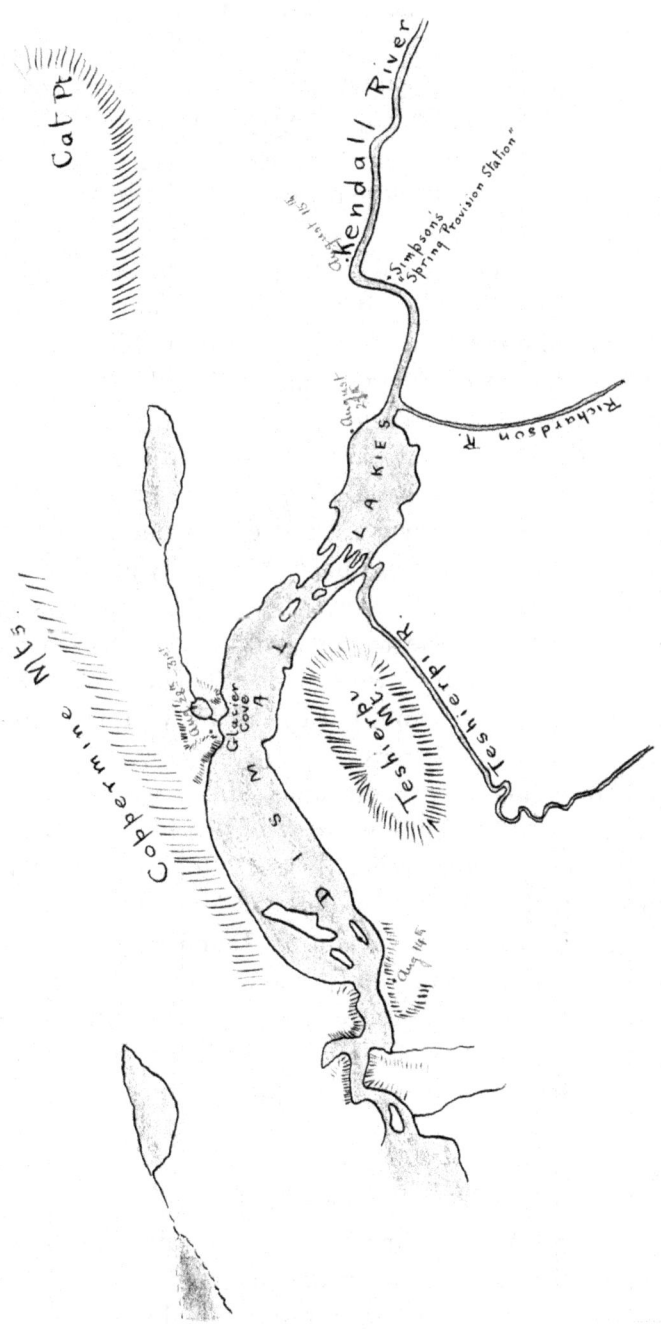

Cat Pt.

Kendal River

"Simpson's
Spring Provision Station"

Richardson R.

August 16

August 24

LAKIES

Teshierpi MES.

Ridge R.

Coppermine Mts.

Glacier Cove

M'SISOO

Aug 16

of the river. Came across a camp, quite extensive remains of a log shack, this must be Simpson's "Spring Camp". I went over to Richardson River nearly, but saw no signs of deer. A most dreary outlook. Back to camp by 7.30, we had fish for breakfast. I caught a 5 1/2 lb. lake trout crossing the stream, a most pretty fish. Took up the net, and broke camp in a rain. Left at 9.30. The river proved unexpectedly difficult, long stretches of boulder-strewn rapids, very difficult to steer down. About 10.30 or 11, we landed and climbed a rather high limestone hill (220 ft. by aneroid). We could see the lower Dismal Lake quite clearly, and several other lakes. On again, a very trying job, frequently had to jump out and wade, so had the Doctor. We had a meagre lunch of hardtack, then on again. The river was of much the same character throughout. It rained all day, and very heavily in the afternoon. Some very bad parts lower down, bad runs, it will be a hard job to get back. We finally got to the cañon at 5.05, we landed and went ahead to investigate. The water is very swift and broke—it runs close to the rocky wall at one place—a bad "set" but I thought I could manage it so we started off. The broken water at the top the Polaris came through fine, but it was difficult work. Above the "set" into the wall she struck a rock, which threw her off her course, only a desperate effort saved her from being dashed against a wall. She had a bad hole knocked in her bow, so we had to land at once in a sinking condition. We made camp right there, a decidedly pretty spot, lots of blueberries on larger bushes, and lots of fairly large spruce. We lit a fire in a pouring rain but it cleared up by 7, and we had a comfortable tea on the fish I caught. It came out finer p.m., a streak of sunlight on the opposite shore made the spot look very pretty. Went to bed about ten, Sandberg washing clothes till late. It froze tonight, the wall of the tent quite stiff with ice.

**Thursday, August 17th.** Awoke at 6.30 to find it raining, a rainy misty morning. Got up at 7, much depressed at outlook and reaction after yesterday's strain. We had breakfast in the rain, then got the canoe up to the fire. Sandberg worked on the net while I worked on the canoe, repairing her. It come out finer, but occasional showers. Out of sorts all

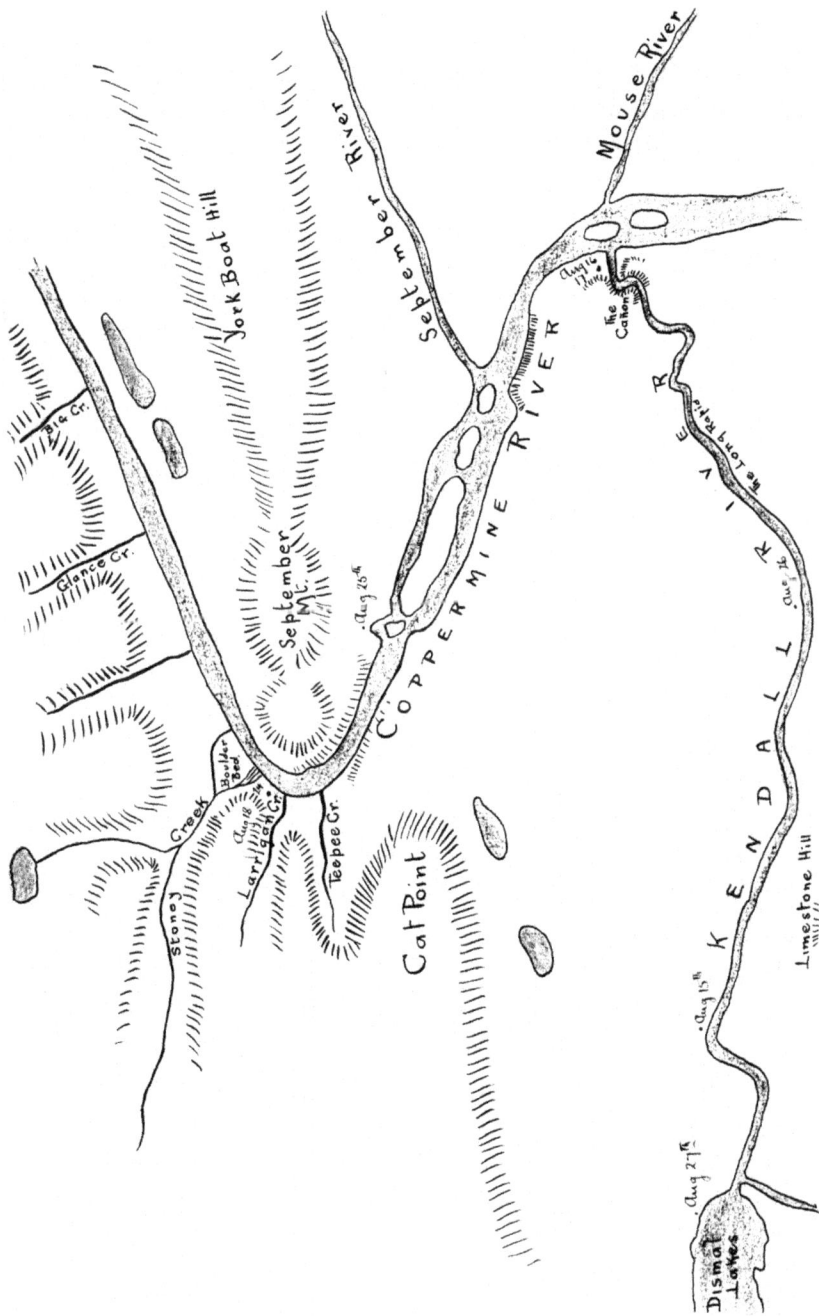

Mouse River

September River

York Boat Hill

September Mt.

COPPERMINE RIVER

The Canon
Aug 12th

Big Cr.

Glance Cr.

Boulder Bed
Aug 18th
Larrigan Cr.
Teepee Cr.
Stoney Creek

Calf Point

Aug 25th

KENDALL RIVER
Be Long River
Aug 16th

Limestone Hill

Aug 15th

Aug 27th

Dismal Lakes

day, mental mostly, this is a strain on one. We decided to stay where we are, so I took out our things and dried them, then to make a picture of the ravine. Then the Doctor and I walked down the river. We walked on to the third ridge and from here, about three miles from our camp, we got quite an extensive view. Above the hill the river runs NW compass, below N 85°W. Opposite the ridge are many islands, one of them quite large. Very raw and cold on hill, we walked back to camp in 1 hr. 15 min. walking hard, say three miles. Shot some ptarmigan on our way, and a good stew for tea. A cloudy night, occasional rifts of blue. Saw rather old moon due north at ten. Coldish, 37°.

**Friday, August 18th.** Fine bright breezy morning, wind about due S true. and quite a strong breeze. 44° at 8 a.m. Took up net, nothing in it. Breakfast: potatoes and hominy fried, then broke camp. Made a cache of some biscuit and 6 Ibs of tobacco, etc. Made a start at 9.15, a lovely bright morning. The river was very swift, I should judge about 5 or 6 miles. It was shallow, we took 28 minutes to pass the long island. The banks got, higher and fairly well wooded. At 10.53 we passed what we took to be the small creek and landed, We decided to go a little further down the river, went on about 2 miles landing again at 11.17. Made camp at a rather pretty spot, shot some ptarmigan and set out the net, at once it got in an awful tangle. We had a late lunch after getting things fixed up, then started out to explore. The Doctor followed the river along, I went up a kind of stoney draw with ice at the top, followed over a level stoney flat, and struck the creek. Crossed this and went up the side of the mountain. About half way up I sighted a bull caribou, and after patiently stalking it, I got within 200 yards. He was traveling west as fast as I could, so I opened fire. At the first shot he ran around in a circle. I fired 3 times without effect, the 4th shot brought him down. I found my first shot had taken out one eye, one had broken his horn, and the last had hit him in the back. This was a great piece of fortune. I butchered him carefully, it was showery at the time with fine spells, and I packed the back and one leg into camp, about 75 or 80 lbs at least. Made fire, the Doctor showed up presently, and we had ptarmigan and caribou for tea.

A cold rainy night, tried to sharpen axe and knives but it was too dark. Boiled some meat, cleaned the guns, and to bed about 10.30. At 9.30 Temp. 39°, raining hard.

**Saturday, August 19th.** A cloudy morning, turning finer in the course of the day. Up at 7 and a breakfast of caribou, then we went out and brought in the rest of the meat. I felt very tired and out of sorts. An early lunch, then surveying around the mouth of Stoney Creek and down the river. Back to camp by 6, and tea, fixed up meat, and to bed early.

**Sunday, August 20th.** A fine bright frosty morning. I was up at 2.30, a brilliant sky, two planets in conjunction, and old moon to north. At 5.30 we got up, the ground had a heavy frost and 1/4" ice in pail. Breakfast, then started at 7.40. Proceeded S 41°W for 1 mile 160 yds, we crossed the creek and went up it about 3/4 mile, direction N 45°W compass. We took to the mountains on north side then, and went up them about N 45°E along the side of an arroyo. Crossed to the top of the ridge, and found one ridge succeeding another of the same kind of rock [these are the Coppermine River lavas, which there form a northerly-dipping set of "stairs" due to erosion of individual lava flows]. Elevation of summit 1170 ft. and the country remained at more or less this elevation. We had lunch on the other side of a saddle, then turned to the south, and followed to the east of the two branches followed this to the junction, then went up the W branch, N bank about 1 1/2 miles. Then back along the creek searching for copper float, finding none. Back to camp, tired. A good tea of caribou and coffee.

**Monday, August 21st.** Raining hard from about 2 till about 6 a.m., a cloudy threatening morning when we got up, the mountains hidden by mist. It turned fine and strong west wind, cold: 39° at 7 a.m, 42° at 7 p.m. We left camp at 8.05, and went along the top of the south branch of the mountains. A cold bleak job and a very disappointing one. We had a very meagre lunch on the ridge, then descended to the south branch of Stoney Creek. Went up this right to the divide, saw three deer there.

A most bleak and uninviting prospect, much the same character as the view to the NNW. This was about due west. Back to camp, very tired both of us, we had walked over twenty miles today.

**Tuesday, August 22ⁿᵈ.** Awake at 2.30, 5.00, and 7.00, raining hard all the time. Got up at 7.30, a fierce day, pouring rain and west wind. We stayed close to camp all morning, heavy rain and mist made any field work impossible. It cleared up a little at p.m., still continued to rain though until about five, when it stopped and showed a few signs of clearing. Along the river bank found some conglomerate. To bed early.

**Wednesday, August 23ᵗʰ.** A dull morning, mountains hidden by mist at first, but it cleared up gradually in the forenoon, and came out a fine afternoon and a lovely clear evening. We got up early, 5.30, and were started down the river by 7.10 to explore that side. Found some more conglomerate below Stoney Creek, I found it quite cold in the morning. Walked along, and came to another small creek that we called "Glance Creek" because we found some glance float there. Then on, and at noon came to quite a large creek that we called "Big Creek". We could not ford it at that point, so after having lunch followed it up for a mile, and then crossed to the hills on the other side The river makes a big bend here, we got quite an extensive view. Then down to the river below the elbow past a deep little lake among shingle hills of glacial drift, and struck the river. A fine tracking bank here, like a cobble paved street. We struck the river about half a mile below Big Creek and about 3 miles below Glance Creek. I was quite tired with my clumsy wet waders, and it was quite warm bright and clear: no flies. We walked up the high well–wooded bank of the river, got to camp about 5.30, both tired. A wash in the river, taking advantage of the hot sun, then tea, caribou, rice and raisins. A clear cold night, 34° at 8.30.

**Thursday, August 24ᵗʰ.** A fine bright morning, heavy frost, mist on the river early. We got up at 5.30 and breakfast under way down to Glance Greek at 8. It was certainly lovely weather, quite hot, no flies.

We walked along through the spruce woods on the top of the rocky bank most of the way. Took a few pictures at boulder bed, etc., looking down the river. We struck Glance Creek, about a mile from the river,

August Sandberg in the Coppermine Mountains

and followed it up finding much copper glance float, and presently a large mass of it that looked like *in situ* to me. We looked around here a while and presently went further up leaving it doubtful, Presently came to an even larger mass of it, no doubt about this being in place. We followed this creek nearly to the top, but no more glance nor copper signs, so we went to a spur on the mountains and displayed our flags in honour of our discoveries. Lunch by the creek, then over the hill to look at the next creek, between Glance and Stoney, but nothing to see there. Back to the camp by four p.m., a lovely afternoon. Overhauled

George Douglas in the Coppermine Mountains

and dried some of our things and had leisurely tea. The river looked lovely tonight, it was a fine day all through, but signs of rain at sunset. At 7.10 the Dr. boiled a thermometer, it gave 98.7°. Bar. 29.29, T. 45°.

**Friday, August 25th.** A lovely morning, turning cloudy but beautiful up to 8. We got up at 5.10, broke camp, and made a start at 8.40 tracking. We got on fairly well, passed the spot where we had landed the other day at 9.20, and got to the end of the island at 11. It was a perfect day, fine and clear, a few small cumulus clouds, no wind, except a few puffs from the south. The land fairly bathed in sunshine. We decided to camp at this end of the island, where the mountains make their closest approach to the river, so paddled across to the east side and made our camp in a pleasant little corner. Had a hot lunch, tea and caribou steak, and started at 12.53 due north compass. Struck the mountains to find limestone at the base, climbed further up and got trap rock again. From, this point we could get a fine view of the valley, looking south up the river. The island appears to be about 3 miles long. A fairly large stream comes in from the NE about the same flow as the Kendall R. but much larger valley. This is a discovery, I christened the river the "September River" We went further east along the ridge and found conglomerate again. We went up to the top of the hill, 678 ft. elev., took some bearings and sat down for a while–a glorious afternoon. Back to our camp and a tea deluxe, bread—first time for 16 days, caribou steak, soup, rice and raisins, coffee. Both filled up to full extent! Took rifle and walked to top of hill behind camp. A lovely evening, like Arizona, same tints to a clear sky, the mountains without trees reminded me strongly. The glow on eastern hills and purple haze in east—all Arizona. To tent about 8 and turned in. Bearings from top of September Hill: Elevation 680 ft., September River comes in from east general direction. Visible about 6 or 8 miles above island N 80°E. Mouth of Kendall River S 42°E. Mouse River comes in on other side of island a little above Kendall R. Limestone hill on Kendall R. bears 27°W, (Identity of this not certain but looks much like it (it was)). General direction of Coppermine river from top to bottom of Island N 85°E. It then bends (going up) to S 45°E, Glance Creek: N 12°W, Kendell Mts. bear S 42°W. The high bluff at S end: S 34°W.

**Saturday, August 26th.** Another lovely fine bright frosty morning, Temp. 30° at 7, probably lower early as the ground was slightly frozen. We got up at 5.20 and breakfast, broke our comfortable camp after another big feed, and on our way by 7.45. Paddled across and started tracking a little above the lower end of island at 8.07. Fresh SE wind. It was bad tracking in places, and I had to put on my waders and wade the boat up in places. Higher up it got better and we passed the end of it at 9.25. At the limestone bluff I had to cross over to a little gravel island and wade the boat up, then back again to the west side of river. We got to the Kendall R. and our old camping place at 11.07. Had lunch, picked up our cache, re-stowed the boat, and made a start up the cañon at 12.20. It was a hard pull, and deep swift water, but we managed to take it all right. Then the next swift water at the curve, and a fairly long stretch of paddling. The river looked very lovely, the autumn tints on grass and shrubs were very fine and most of the afternoon we had fine sun. It blew hard from the SSE, true, and clouded up about 3, a very pronounced halo, and again at 5. The long stoney stretch was hell, but we got up all right. I waded the boat nearly the whole time, Sandberg helping me with a pull on the painter at times. We kept at it till about 6, then began to look for a camping place and turned down one place at the bottom end of a long stoney stretch. I waded her up this, dead tired, and we camped above it on the north side, a hill, but good place for tent. A big feed of caribou.

**Sunday, August 27<sup>th</sup>.** Dull windy morning, a little blue showing in E, heavy black to N and NW, a regular winter sky. Got up at 5.10 fire and breakfast of caribou steak and hardtack and broke camp. Got under weigh [way] at 7.35 a.m. A quiet stretch, then above the first rapids I saw a deer. Shot at it from canoe, it ran a while evidently wounded. I landed and found it dying. It was a doe, we took the hindquarters, liver, heart, tongue and left the rest. We got on with fair success, frequent stretches of wading and a few of paddling. A lovely day, sky like an Atlantic autumn sky, marble, only a streak of blue here and there. The autumn tints along the shore very beautiful, black flies very bad, We got to Limestone Ridge at 11.40, and walked to the top. Took the following bearings: narrows between 1<sup>rst</sup> and 2<sup>nd</sup> Dismal Lake: N 76°W; September Hill: N 27°E; Coppermine Mtns: high point to N. Stoney Cr. N 14°E; high point to S. Stoney Cr: N 7°E; Sept. Mtns. between N 20°E and N 46°E, distant about 15 miles, General course of Kendall as far as traceable to east: N 60°E. We had lunch down by the boat, and got underway again at 1.20. It was a lovely afternoon, came out fine bright and warm, the traveling was easier in places, but still mostly wading. We passed our old camp without my noticing the place, it was only when I saw Simpson's old camp that I realized where we were at. I landed and took a photo of this, then a last wade up stream, and a paddle against swift water and we came out over the sandy bar at 5.10 p.m. A lovely afternoon, and the lake looked very beautiful with the strong sun on the mountains. We landed a short distance along the shore and made camp. I was very fatigued, and the black flies were fierce, the worst yet. A bad place for the tent, but we put it up anyhow, and cut lots of spruce and a comfortable tea of liver and steak, rice and apples – lots of everything, with a pipe of black tobacco to finish up with, and to bed early. A lovely night, reflections of Kendall Mts and Teshierpi very lovely.

**Monday, August 28ᵗʰ.** Awake at 4.30 and out to have a look around. I thought it was raining at first but it was only a heavy mist, the sun was rising in a clear sky. Back to bed and slept until 7.30, the record for late rising on this trip, but not so much need for early rising now. A leisurely, breakfast, lovely day with fresh east wind. I felt a great relief to look at that quiet lake instead of heavy rapids ahead. Left at 10.27, entered narrows past first point at 11.20. We went up what <u>second</u> thoughts indicated the channel, and found we were going up quite a large river, with the flow larger than the Dease. We christened this the Teshierpi River. Paddled along the shore of the second lake till 12.45, when we landed and had lunch on a sandy beach, then climbed Teshierpi Mountain. Altitude 780 ft., distance from canoe to first summit 1 mile 1120 yds. Took following bearings from second summit. Altitude the same practically. Bearings from 2ⁿᵈ summit Teshierpi Mt. #3 Dismal Lake lies E and W between Second narrows and Dismal Point. 2ⁿᵈ narrows. S.88°W; Dismal Point. S 85°W. There is a large lake to South lying between S 34°W and S 22°W, distance 10 m (This was Mountain Lake). To the east, bearing S 13°W, is a high mountain—highest point visible to south (Dorothy and Marian). Between lake and 2nd narrows country covered with lakes, fairly low, 2nd Dismal Lake lies about WSW and ENE, but is very irregular in shape. 3rd Dismal Lake very irregular, and a long bay runs in to the NE corner, and a long lake continues from there, Divide lies behind Dismal Point and cannot be seen. 1rst summit bears N 71.5°E distance 1360 yds. Bearings from 1rst summit. Kendall R. mouth: S 83°E; 1rst narrows: N 85°E; 1rst Dismal lies S 75°E, N 75°W; Limestone ridge: S 82°E; large lake to north between N 60°E and N 40°E; river runs from that into 2nd Dismal L., its mouth: N 25°E; Teshierpi River runs generally east and west near the lake. Back to our canoe and across the lake, it was quite rough and advisable for me to do the paddling alone. We landed in a little cove of glacial moraine with a pure gravel beach and a flat spot between two gravel hills, a fine place for a camp. Made camp leisurely and a leisurely tea of potatoes, soup, hardtack, liver and steak. A dull stormy looking evening, high E wind.

**Tuesday, August 29th.** Awake at 4.30, raining hard, and raining still at 5.30 and 6.00. Lay awake this morning for a short time, then a sleep and up at 7.30. Dull stormy morning, Teshierpi Mtn hidden by mist, strong NE (true) wind. Leisurely breakfast. Lunch Point bears N 260°W, S 85°W. We walked around to look at what I had taken from Teshierpi Hill to be an old camp, and sure enough, signs of camping, about the same age as all the others that I have taken to be Simpson's and the same character chopping. The shore line to all moraine gravel hills from 70-100 ft. steep, lakes between them. We had to go back to our camp, and climbed the ridge from there. It came down thick mist, we had to steer back by compass. This put a stop to further explanation. We kept close to camp all day. A leisurely lunch, made bread, and then stayed in tent. Fierce weather, dense mist, high N wind, a regular gale, fortunate we are in a sheltered spot. We put in the time reading, sharpening knives, etc., and played euchre and stud poker, then made an elaborate tea. Soup, tenderloin steak, tongue, bread, rice and raisins, and tea, all the best yet. A wild night, a heavy sea running on lake, thick mist, gale about 40-50 miles from north. Glad indeed to have the snug camp we are in. Temp. 42°, Bar. 28.5. Turned in early, heavy rain as well as mist and gale as I wrote this up.

**Wednesday, August 30th.** Awake several times early, always to find it raining hard. We finally got up and made breakfast at 8, a rainy morning, heavy showers till noon. Stayed in tent, tried to make map but gave it up at 12. It showed some signs of clearing, we had lunch and made a start over the hill. Rainy, then heavy mist. Lots more conglomerate, epidote, amygdaloid, etc. Too misty to go any further so back to camp. Set out the net, then to make a new fireplace, a fine big log. Bread, potatoes, boiled ptarmigan, fried loin, we are getting great epicures. A dark misty night, less wind, and it was more to NE.

**Tuesday, August 31st.** Up early, drizzling rain at 5. To sleep again till 7.30 when we got up. A drizzling misty morning. We went to the net after lighting the fire, but there was nothing in it. I caught a small lake trout trolling, we had this boiled for breakfast with potatoes and soup. We made a start about 9 to explore the mountains NE, but it came on very wet, and so misty we could see nothing, so after wandering around in a circle we made our way back to camp, and made the best of things. It stopped raining about 2, but remained very thick until about 5.00, when the mist began to lift, and the mountains to show up. It came out a fairly clear evening, mare's tails NW to SE. We put in the afternoon Sandberg washing, self fixing axe, steak, rice & apples, coffee. Temp 43°. Lake 48.

**Friday, September 1st.** Dull hazy morning, mountain tops covered with mist, mirage on lake. Slept badly, we got up at 3.30 and breakfast, and under way at 6.05. Passed our old camp at 7.30, and landed a little beyond to look at the eskimo's camp. He had been there since we were, but now no sign of him around. I walked up in bare feet, it is now getting quite cold. We went on at 8, and passed into 3rd lake at 8.30-35, turning perceptibly colder. A short distance up the lake I saw a deer lying down. I went up, it started to run. I shot at 340 yards, also her calf a little further. We landed, took the hindquarters, and on again. A lovely morning, bright, and the mist lifting from the mountains. The south shore of the lake is deeply indented with bays, and two islands. On one of those we saw a bull caribou, a fine big one and a dead easy shot against the sky, but we had more meat already than we could pack, and time was more precious, as it began to cloud up and to blow. I was afraid of another gale out of the north, which would have been very bad for us. I made the traverse to Dismal Point before lunch, we landed then on the east side of it, and a miserable lunch of cold meat that was no good. Feet cold going with bare feet. The wind increased, blowing from the west, and pretty hard. Beyond Dismal Point the sea began to run heavy, and as we rounded to south it became worse. A heavy rain squall to add to our discomfort, we were then at the point where we had had lunch on our way in. It was quite a struggle to got to a little sheltered bay to the west of this point, we waited there awhile, it looked like a camp, but such a miserable prospect that I decided to push on at any cost. A hard struggle, but the wind began to drop, and though it still blew pretty fresh we were able to make headway. Things improved as we went on. We halted at 3.12 p.m. A dull windy afternoon, as we paddled up the lake the lower end seemed interminable to us in our tired condition, but we finally got to our old camping place at 6.53 p.m. having been at it for nearly 13 hours. Made camp, same old place and to bed quickly as we could. A cold night, but not freezing.

**Saturday, September 2nd.** A lovely at 4.30, misty, but blue sky above. I was out then, but turned in again and slept till 7.30. It was then dull and windy, and the tent dripping. We got up, breakfast. and fixed up our stuff in shape for packing over the divide. Took some of it and the canoe to the top of the hill, then back to the tent for lunch. Made a start at 2 p.m. Teshierpi L. lies a little to east of south. It came on rainy, we packed the stuff to the little pond at the end of the first lake and thought to make camp there, but it cleared off a little, so we went ahead, brought the canoe over and took all the stuff to the divide between the first and second ponds and made a comfortable camp in a thick clump of spruce, a treat to get where there is good firewood. A cold windy evening: 33° at 8, sometimes patches of clear sky, sometimes heavily overcast. Made some tent pegs and turned in at 9.30.

**Sunday, September 3ʳᵈ.** Awake at 2.30? too dark to see watch. It was then cold and had been snowing. To bed again, and up at 7.30, the ground covered with snow, and strong south wind. Breakfast and got our things fixed up in a heavy snow. Finally made a start at 8.40, although it was then snowing. We took to the west side of the ravine, went ahead, and packed all the stuff about 3/4 of a mile, leaving the canoe. There was a good crossing place, and better walking and direction on the E side so I went back and got the canoe. Waded, pushed, carried, and paddled her up single handed and barefooted, bitterly cold, the lakes frozen in places, and I found that the swamp is frozen permanently only a foot or so below the surface. We passed the point, on which we had camped, at 12, took all our stuff down to the pond and had lunch under the lee of the canoe. It turned finer p.m. but still occasional showers. We crossed the divide, and I found a good camping place near the big pond on the other side, so we brought all our stuff along, coming to camp at 3.45. A cold windy evening, turning brighter. Climbed the hill behind our camp to get some bearings, a cold windy job. Back to our camp, a snug spot, fine floor to our tent and lots of wood. A comfortable tea, erbswurst, steaks, and rice, and lots of it. Saw 7 deer on plain on the other side, T. 31° at 8, Strong N wind, clouding up to west. From Divide Hill: Sandy Creek hill bears S 5°E, ?Granite Ridge between S 26°E and S 20°E.

**Monday, September 4ᵗʰ.** Up at 4.30, a cold morning and some clear sky. To bed again till 6.30, then we had breakfast. It was 29° at 7 a.m. and I was cold last night. A lovely fine morning, slight SE wind. Broke camp, and under way at 8.15. Ice on some of the small ponds. Got to the crest of the hill above Sandy Creek at 11.30. Hanbury's Kopje: S 41°E, ?15 miles; Granite Ridge between S 60°E and S 80°E; Sandy Creek Camp, S.

Simpson at S 28°E and S 32°E; Dease Mtn: S 42°E. Landed at Sandy Creek a little above, I thought below, as we headed across, our old camp, at 12 and got to the little rapid at 1.40 where we had lunch. It was dullish, not as cold, we waded the boat down barefoot. The creek is just about the same height as when we were here last. Cold wading. Paddled down the steep winding strata above Sandy Creek Hill, very enjoyable, the colors very beautiful, especially on Sandy Creek itself....dark rich red of small heaths and huckleberry bushes, and bright yellow of willows. The latter are losing their leaves fast. We attempted to go on beyond the hill, but it was too shallow, so turned back to our old portage landing, and each took a load clear across to just opposite our camp of August 11th. Sandberg took two more loads, while I took the canoe and one load around. Easier than going up but had to wade in places. Water perceptibly warmer than above. Decided to move around the next bend, so Sandberg brought part of the stuff while I took two loads in the canoe, and we made camp at the last of the spruce down the river, a pretty spot and a comfortable camp. Came to camp at 4.50.

**Tuesday, September 5th.** A lovely morning, sunny and warmer. 35° at 7. Got up at 5.20, breakfast and broke camp, and took our stuff a little below camp to a good landing place, and got underway at 8. S took his pack and walked, I took the canoe, it was necessary to wade a good deal and to lighten the canoe at a couple of places, but we made good headway. I was barefoot, the water was just freezing, so it was a rather cold proposition but luckily the day was not cold. We got to the rocky portage sometime about 10.30, carried the stuff over, also the canoe, then on to our camp at Simpson's Point at 12 noon. A dullish day, faint sun, enough to throw shadow. We had a hot lunch, erbswurst and deer steak. I saw a bunch of deer to W after and we decided to hunt as we began to need the meat. Sandberg took to R and I took to L and we circled towards them. I got to within 300 yds, it was all open country and they saw me, no shelter, so I opened fire, shot wild the first five shots then wounded one The others ran toward S. I fired a couple of shots as they ran, then turned my attention to the wounded deer, fired three times more at it as it walked slowly

around. It dropped at the last shot, I found it shot three times through
the guts and once through the shoulder. We took the hindquarters, a
fine young buck, and started to camp. Saw another on our way, fired both
of us at it several times. I thought I hit it twice, but it got away, and we
went back to camp. I was personally much disgusted at the generally
poor shooting especially the last. And unnecessary, as we had already as
much meat as we could pack. We got back to camp at 4.35 and started off
with the canoe at once. Used only the two upper stretches, then dragged
her over the brush, and got to the end of the portage at 5.35. Walked
back fast to camp. Made a sumptuous tea, bread, calf steak and liver, rice
and apples, coffee and malted milk. A lovely evening, mild and quiet.
35°. Bearings from Simpson's Point: Polaris Mountain: N 34°W; Procyon
Mountain: N 42°W; Jupiter Mtn between N 34°E and N 12°W; Granite
Ridge between S 77°E and N 65°E; Hanbury's Kopje: S 42°E; Simpson Mt:
S 23°E; Dease Mt: S 47°E.

**Wednesday, September 6th.** Awake early and out, a bright starry sky,
quite a novelty to see the stars again. Got up at 5.30 and breakfast, a lovely
quiet morning, faint sun, mackerel sky and mild. 39° at 6. A few black
flies around even. Packed our stuff, saw 10 deer to north and a couple of
bulls to south. Left camp at 8 a.m. A fine bright and mild morning, lovely
autumn weather. The dwarf birch brown and crimson, the blueberries still
hanging withered on their bushes. We completed the portage at 12.20, 4
hours and 20 min of very hard work. Had lunch, then started out at 1.25.
It was a lovely afternoon, I took the Polaris while the Dr. took his pack
and walked. The river was about the same as when we came up. I had
to wade occasionally. We made good time, and I found it very pleasant
floating down this little stream on such a lovely afternoon. I noticed
footprints, some small, along the bank. At the junction of Granite Ridge
Creek I landed to get our cache. I found our lard pail gone, and 3 arrows
and a pair of slippers left. The bag had also been taken down. There were
signs of quite extensive camping done here and just below our old camp,
on the high bluff were signs of a big camp, evidently about two weeks
old. Loaded up the cache and down the stream, this is higher than when

we were here last. The Dr. was awaiting me at the junction point with the Dease R. and we went up the river to just above our old "beach" camp. Camped on the other side, a fairly good spot. A disorganized camp, I was very tired. It turned dull and threatening, clouding up in the north, but still warm. A few black flies around. A shower PM then a dull quiet evening. To bed early.

**Thursday, September 7th.** Dull raw morning, strong north wind, overcast, occasional skiffs of snow. Up at 7.30, a leisurely breakfast, oatmeal, beans, caribou loin, then made a start for Hanbury's Kopje. We took some bearings there then walked along the ridge to the Eskimo camp. They had evidently gone for good. We struck across then for Granite Ridge, cold and raw, strong N wind. Got to the east spur and took some bearings from the shelter of a big rock, but it was a very cold job. We could see smoke in the valley below the last camp we made on the East Branch. We thought this might be the Eskimos, I was rather doubtful myself, too big a fire for them (Douglas note: we later found that this was Father Rouvier and Hornby on their way up to Lake Rouvier for the first time). It was too far to investigate, so we turned back and had a brisk walk back to our camp, cold and windy, the ground covered with lots of frozen blueberries. A regular east Canada late November day. Got back to camp at 4.30, an early tea. A cold night, 29° at 8. I plotted some of our bearings, then made hot Teshierpi toddy and to bed. Bearings from Hanbury's Kopje, NE side: Simpson Mountain: S 6°E; Dease Mountain S 35°E; Procyon: N 44°W; Polaris: N 36°W; Jupiter between N 28°W and N 18°E; Caribou Knob: N 72°W; The divide we were heading for on the east branch: N 57°E. Bearings from Granite Ridge: Long Lake (Lake Rouvier): N 60°E, 8 miles? Spion Kop: S 85°E, 2 miles? Big Lake (This was the lake at the head of East river): S, 9 miles? Storm Crest: N 50°E, 28 miles? Sandberg Gulch: N 80°E, 8 miles? Country between Long L. and Big L. has many small lakes.

**Friday, September 8th.** Up at 4.30 but thought this was too early, so slept until 6. A promising morning early, but it came on to snow. The ground was frozen and water taken from the river was skimmed with ice a few minutes later. Beans and caribou for breakfast, then broke camp. Start 8.22, passed Sandy Creek 8.29. Dull, raw and cold, NNW wind. Passed Muddy Creek at 9.33. Landed a little below Muddy Creek, Hanbury's Kopje N 40°E. We struck rapids at 11.30, the Doctor got out with his pack and walked, we got to Notman's Dyke at 12.40. Took off our stuff and had lunch in front of a fire, it was bitterly cold. Climbed the ridge after lunch and took the following bearings: Notman's dyke runs N 56°W; Caribou Bluff: N 40°W; Procyon: N 27°E; End of Granite Ridge: N 21°E; Jupiter between N 11°W and N 11°E; Red Bluff: S 60°W; Simpson Mountain: N 73°E. We made a start at 2.25, Sandberg walked to the shallow elbow and I took the canoe down. I had to portage the stuff there. We both paddled then and made good headway all afternoon, the river was of good character, such rapids (excepting one that we had to portage part of the stuff at) we could both run, so we made good time till we got to the rock ledges. The river was shallow here, we tried to run but stuck, I landed Sandberg and walked the canoe down. Then came better water. We passed our camp at the island at 6.30 and camped just below on the other side, a rather good spot. Bread and potatoes and caribou stew, it was dark by the time we had supper. The run down was very pleasant this afternoon, it was dead calm, and autumn is now far advanced. The willows have lost most of their leaves, such as remain are yellow, there is far less red in the colours here than around the divide. Saw a few duck, and a regular army of geese flying SE. It was cold, 32° at 7 p.m. and turned a rather cold evening. A glow on clouds at sunset.

**Saturday, September 9th.** Dull overcast, fresh N wind, cold 28° at 7.00. Up at 4.30, but to bed again, and slept till 6.30, when we had breakfast and broke camp. Started 9.10. The old island was just below our camp, and swift water until 10. I walked twice. At 10 we struck quiet water again. Then more rapids, quite exciting to run, then a quiet stretch. Came to Clay Elbow at 11, the place where we landed to look on our way up. Got to our lunch place at 12.02. Stopped for lunch at 12.30 on a high bank just below the East River, made fire, warmed our steaks and

had chocolate. Underway again at 1.40, passed Goose Reach Point at
2.32. A pleasant paddle down the river though it was a dull, cold, cloudy
afternoon. We got to our old camp at Observation Hill at 4.00, went on
a little further and camped on the opposite side of the river at a good
place. Got things fixed, then went for a walk to the top of Red Bluff.
Saw a dog on the opposite side of the river. A bleak looking prospect. I
thought I could identify Notman Dyke. The hills beyond were covered
with snow. No game in sight. I went back to camp and tea, and turned
in early. Dull, but a lovely glow on Observation Hill opposite, a delicate,
almost translucent, terra cotta.

**Sunday, September 10th.** Dull morning, awake early, head aching,
troubled sleep until 7.30 when we got up. Tried the canoe to find the
leak. Breakfast, beans, and caribou. Started for Observation Hill about
9. Landed on other side, shot a doe, the calf got away. Then to the hill.
Bearings: Notman Dyke, south end about N 9°W; Procyon mountain: N
20°W; Jupiter mountains, between N 13°W and N 2°W; Granite Ridge,
south end: N; Simpson mountain: N 7°W; Bald Knob: S 35°W; Narrakazzae
islands: S 11°W; Big island, center line: S 25°W; Bell's "South Branch"
runs about E and W; it's mouth bears S 22°W. From the North side of
this runs a ridge or long series of ridges about N 10°W and S 10° for
about 12 or 15 miles. There is a low gap with an inconspicuous detached
round mountain bearing N 24°E. To the NE of this is another high ridge,
"Snow Spot Ridge" between N 32°E and N 9°E. This ridge runs apparently
about NW and SE (Later note by Douglas: Think this should be NE &
SW). To the extreme end a high mountain can be seen, N 6°E (Dorothy
and Marian? or Storm Crest?). Back to our canoe and camp. It came out
a lovely afternoon, regular Canadian autumn trying to be summer. We
broke camp and made a start at 2.30 p.m. Very pleasant paddling down to
the South Branch, then I had to wade. I landed the stuff at the same spot
at which we had embarked on our way up, and portaged it all, including
the canoe, at the same place. Then portaged the stuff, and lowered the
canoe at the box. Sandberg landed with his pack here, and I took the
canoe down. A shallow bar that I had to wade her over, then a very trying

few minutes in the centre channel, got a bad jam at the head, then in what came near being serious trouble at the willows. Finally I decided to make a bold run of it, and came through all right, only shipping a little water. We camped on the hill just below there. I made a cache of some stuff. Had a very pleasant supper out on the point, liver, rice and apples. Bright moonlight. Temp. 27°.

**Monday, September 11**[th]. A lovely fine morning, I was out at 2 a.m., a curious twisted light in the sky, I could not say whether cloud or aurora, violent and extraordinary rapid motion in it, a most curious sight (Douglas note: this was my first sight of the very close aurora we sometimes had, and what was the most notable instance of it). Up at 5.20 and breakfast, T 28° at 7. I walked to the top of the hill, but could see nothing of the direction of the camp. Got under way at 7.26. Broke a hole in the canoe, had to land, make a fire, and repair it. Got under way again and passed the place where we broke her before at 9:20, and broke her again at almost the same place, but carried on till we got to our old camp (No 1) where I stopped and repaired her while the Doctor packed the stuff down. A pleasant run down, I took Sandberg's pack below the island, and himself about a half a mile further down. Between the island there is very little paddling, all swift shallow rapids. Had lunch in a spruce grove opposite the grave. A lovely autumn day and a good lunch, tenderloin, apples, rice and biscuit. Took the following from the Indian grave. Bald knob: N 28°E; Observation Hill: N 33°E; Confidence Hill: S 35°W. Left Indian grave at 1.42. Got to Hodgson's Point about 3. I did not note the time. The place looked much barer than I expected, the willows having lost their leaves makes a great difference. The house was finished and looked remarkably snug. We did not see Lion at first, he was busy papering the house inside, and came out as I went up with some of our stuff. He was looking very well, and much bearded, it was fine to see him again. We brought up our stuff in a slovenly way, had too much to talk to each other about. I put up the tent, and got out the "wad-ma-tilt" and cot. We had a comfortable tea and talk after, the house seems a regular little palace to us now.

|  | Took | Cached | Used |
|---|---|---|---|
| Flour | 50 lbs | 9 lbs | 41 lbs |
| Bacon* | 26 lbs | 11 lbs | 15 lbs |
| Sugar | 35 lbs | 3 lbs | 32 lbs |
| Rolled Oats | 10 lbs | 7 lbs | 3 lbs |
| Beans | 16 lbs | 9 lbs | 7 lbs |
| Rice | 17 lbs | 3 lbs | 14 lbs |
| Hominy | 14 lbs | 7 lbs | 7 lbs |
| Hard Tack | 24 lbs | 12 lbs | 12 lbs |
| Erbswurst | 8 lbs | 5 lbs | 3 lbs |
| Dessicated Potatoes | 10 lbs | 4 lbs | 6 lbs |
| Chocolate Peter's | 6 lbs | —— | 6 lbs |
| Chocolate Bakers | 2 lbs | 1/2 lbs | 1 1/2 lbs |
| Cocoa | 1 lb | —— | 1 lb |
| Tea | 3 lbs | —— | 3 lbs |
| Apples | 3 lbs | —— | 3 lbs |
| Raisins | 3 lbs | 1 lb | 2 lbs |
| Raspberries | 3 lbs | 2 lbs | 1 lb |
|  |  |  | 157 lbs |

45 Days
Shot 7 deer of which we used say
meat without bone                                                        170 lbs
Shot 10 young and 1 old geese                                      11 lbs
Shot 26 ptarmigan                                                          5 lbs
Shot 4 ducks                                                                  2 lbs
About 20 lbs fresh fish                                                20 lbs

|  | Meat | Total | 208 lbs |
|---|---|---|---|
| Dry Food |  | Total | 157 lbs |
|  |  | About | 365 lbs |

Or 4 lbs per man day

Note added by Douglas later: I think the figure for Deer meat far too high. I should think 130 lbs at the outside.

* We had some of this bacon left when we got to Ft. Norman next summer. It was some we got for our voyage down the Athabasca when we left in May 1911.

Morenci, Ariz., May 20<sup>th</sup>, 1910

Dear Cousin James,

I am getting a general idea of the country along the Athabasca, Slave and Mackenzie Rivers, and though a lot of information is still lacking, a plan is taking shape.

The nearest R.R. point is Edmonton: from Edmonton to Athabasca Landing, a distance of 100 miles, transportation is by wagon. From Athabasca Landing to Grand Rapids a steamer runs when business offers, or scows as an alternative means of transport. From Grand Rapids to Fort McMurray, ninety miles, is mostly rapids with partly navigable stretches and wagon transport controlled by H.B Co. From Ft. McMurray to Smith's Landing on the Slave River a steamer runs and a portage by wagon for about 18 miles to Fort Smith. From Fort Smith a steamer owned by H.B. Co., runs all the way down the Slave River, Great Slave Lake and Mackenzie River.

I do not know yet whether it would be better to make Fort Rae on the Great Slave Lake our base and starting point or whether there are any posts on Great Bear Lake that would be better.

At present I am inclined to the latter route even if there were no post nearer than Fort Norman at the discharge of the Great Bear River into the Mackenzie R.

| From Fort Rae | From Fort Norman |
|---|---|
| About 250 miles of small lakes and streams scarcely half the distance navigable. This would land us near the headwaters of the Coppermine R. and with about 3 weeks journey and a good many portages before we got where we wanted. | 100 miles swift river to Great Bear Lake. 150 miles of lake 50 miles portage to Coppermine River which would land us close to where we wanted to be. |

In Franklin's first expedition he took the former route although he conceived that the route through Great Bear might be better. In his second expedition, of which I have no account, he established winter quarters on Great Bear Lake though of his route thereafter I know nothing at present.

I am going to write to the Hudson's Bay Co. for information on these routes, supplies carried at their posts. etc.

Beside one person the amount of stuff that would be carried in a 15 ft canoe on such a trip would be limited to say 300lbs and of this about one-half would be tent, sleeping bag, instruments, utensils, arms, ammunition, etc.

Such supplies as can be carried we would have to save for the latter part of the journey over the divide and as everything that summer and perhaps following winter. We still have to depend on living off the country: a prospector will do this to that limit anyway but he won't as a general thing depend on doing it; however in this case it is one of the conditions we must expect to meet.

At present the Great Bear route seems best but I think it important that we should ascend the Great Bear River that autumn. If there is any H.B. post on Great Bear Lake, make our winter camp near it; if not, establish some kind of winter quarters there anyhow and cross the divide before the snow melts in the following spring.

Therefore, it would be better to figure on availing ourselves of steamers and H.B. Co. transport to Fort Norman; these should land us there in the end of July and give us a chance to establish ourselves on Great Bear Lake before the ice formed in September. Of transport by dog team in winter I don't know much; we would probably find it advisable to make some kind of a depot at the nearest point on the Coppermine River during the winter. I expect that country would soon teach us all about dogs as motive power but I would prefer to depend as much as possible on ourselves.

The cost of transport of two men, two canoes and about 1000 lbs of stuff from Edmonton to Fort Norman would come to over $350.00: the distance is about 1500 miles. To Edmonton, from say Toronto, about $200.00

or 250.00. The canoes, which I should have specially built introducing certain features, would cost about $80.00 each. The cost of instruments depends on how elaborate we wanted to be; a comparatively simple outfit would run up to a couple of hundred dollars.

August Sandberg and George Douglas

Clothing and food I have no line on as yet but we may guess the cost of the expedition up to start from Fort Norman to exceed $1800.00 anyway. If it was composed of three members the cost would be nearly a third more.

When we first spoke about this trip I don't know that you realized how expensive it would be. I did not and even now these figures give me a feeling of misgiving and dismay at their extravagance. If we went by canoe from Athabasca Landing to Fort Norman forwarding the bulk of our stuff we might save about $200.00 but this at the probable expense of a year. The waters remain open so short a time in that country.

The cost is certainly going to be considerable and now is the time to realize it. Of course I shall contribute all the available money I can but I fear this won't amount to much when the time comes. For, if when the probable cost is realized and you are still willing to assume what will practically be the entire burden, I shall go East this autumn and start making preparations. All kinds of details must be attended to. I want to give the building of the canoes my supervision and a thoroughly practical trial afloat in swift water and on ice. Everything else about the equipment should be tested out leaving nothing to the chance of its being all right. We should have everything ready on the ground at Athabasca Landing early in May.

Such elaborate attention to detail might be thought hardly necessary but this has not been the experience of successful trips: the more carefully preparations are made and equipment tested out the more sure we are of results.

So when the time comes to start I would not be in shape to contribute much. Now I don't want you to think for a minute I am getting cold feet—on the contrary what I have learned about the country makes me keener than ever to do something up there. But it also makes me realize what we are up against and it would be a great mistake to underestimate this. You must be given some idea of what the cost will be: if you still want to stand for it, I shall be thoroughly glad of such an opportunity and shall certainly do my best.

I should also like to know what you think of a third man to the party; it might be advisable for some things.

So long as we could depend on Dr. Sandberg as geologist I should perhaps not to have any outsiders. Dr. Sandberg left Morenci to get into better shape financially. I think we can rely on him to join; and know that if he does join there will be no quitting on his part.

I don't know whether my brother Lion would care to impact his sea going career on such a proposition but would lay the scheme before him if you saw fit. His experience and training have been such as to make him a desirable man for such work. The only other man I know that I would care to have along is a young Englishman whom I have known in Canada, and the West. He is not a 'scientist' in any way but has other invaluable qualifications of taking anything that comes cheerfully without question and is thoroughly to be depended on.

I would also like to know what you think of writing to the H.B. Co now direct for information. I would define the object of the trip as "geological reconnaissance"—or whether a letter covering the points on which information was required and sent through you to any of their men you might know.

With best regards to all.
Yours always affectionately,
George M. Douglas.

Northcote Farm
Lakefield, Ontario, Canada
Sept 17th, 1910

My dear Doctor,

I don't know what you will think of me for not writing but I had a good deal of trouble with my throat since seeing you. Finally had operations made that seem to have fixed things all right. My health wasn't good and I spent the summer recuperating here with beneficial results.

I dislike the thought of going W. again, I want to get into the best shape for our trip.

I wish you would join me here now or as soon as you can. I have a great deal to tell you about and have collected quite a lot of books on the subject.

I am going to stay on here I think and hope you will write or can come yourself and get a taste of a climate that will fit you better for the North. J. Douglas will probably be in Morenci soon. I hope you will see him. Shall write fully when I hear from you or that you will make this unnecessary by coming yourself.

Yours sincerely,
George M. Douglas

Morenci, Arizona
Oct 27th, 1910

My dear Lion,

Since writing last I made a trip down south from here to meet Dr. Sandberg. He seems to think it doubtful that he will be able to get away this year, or rather, next spring: between ourselves, he is in rather a hole with various unprofitable mining deals and has financial obligations that he must clear off. He will know just how he stands by New Year.

This is upsetting to my plans as the expedition practically depends on his accompanying us.

It leaves me with three alternatives.

(1) Postpone the start until the spring of 1912

(2) Get someone else from the Canadian Geological Survey or elsewhere to attend to geological work

(3) Make a start anyway even by myself should you not care to accompany me under the circumstances and let Dr. Sandberg join us next year.

Taking (1) I am decidedly adverse to losing a year on this proposition. In the second case we would be taking certain risks in getting mixed up with anyone from the Canadian Geological Survey. There would be conflict of authority or might be and starting out on such a trip as this is likely to be, it is a good thing to understand each other thoroughly. A stranger might be a good fellow and we might get on all right or be a good fellow and we would get on badly in spite of it, or might be inclined to kick or make trouble and there would be hell all around. There is a difference between a man coming with Dr. S. and myself to look after a certain well–defined part of the geological work, and a man coming to attend entirely to that. The third course is the one I am most inclined to favour at present. Whether James Douglas would care to stand for the expense of such a trip when the geological results would of necessity be so meagre the first year and all that he would get for his money would be our experience of the country and what not to do (something that often costs a lot to find out) I don't know. I haven't written to him on the subject since I saw Dr. S and write to you now so that you get this much anyway before you sail.

Supposing that he showed any unwillingness to do this, I think we might even make a trip up there on our own. I could stand for canoes and most of the equipment. We could make a flying trip that summer and even be back before the freeze up. Our transportation to Ft. Norman from Edmonton with canoes and stuff would come to about $350.00 and the trip back without canoes etc to about $200.00 or so.

We would do this inside of five months so you would not have to bother about extra leave and it would be a rather fine way of spending a vacation. I have even ordered 2 16' canoes specially built, so that we shall be prepared for it.

Of course it shall be decided by what you and James Douglas think, meantime I hope Sandberg will get his affairs straightened out so he can come.

Hoping to hear from you before you sail.

Always your loving brother,
George M. Douglas

Bisbee, Ariz
Nov 25th, 1910

Dear Cousin James,

Your letters "received", both the copy of that which you wrote to Uncle Archie and his reply. Please excuse my delay in acknowledging the former.

I saw Dr. Sandberg about a month ago, he is doubtful whether he can get away this spring. He has certain financial obligations to fulfill and does not know whether he will be in the clear by next April. He feels that he should prepare himself in certain ways for this trip and doubts whether he can do so in the time. He realizes the difference between a prospecting trip for copper and a geological survey and thinks however well fitted we might be for the former we are neither of us capable of the latter and for that reason another geologist would, in any case, be a good thing! He feels that if you are putting up the money you ought to get the best results. His preference in the matter would be to go on his own resources which he thinks he could do if the trip were to be postponed for a year.

It is not a subject for persuasion or argument as his arguments are unanswerable. Some of them apply as much to myself as to him. It holds me in some perplexity as I have always considered the Dr. to be an essential member.

It leaves us with practically three alternatives.

(1) Make arrangements for someone else to go to take charge of the geological work and if Dr. Sandberg finds himself in a position to accompany us by the time we got ready to start, so much the better.

(2) Make the trip to the country between the Great Bear and the Coppermine R's solamente if no one would come with me, and at least familiarize myself with the locality, people, and methods and be in good shape to at start out with the Dr. when he joined me the following summer.

(3) Postpone the trip for a year

The first of the alternatives depends so much on the man who would come to take charge of the geological work. If some arrangement were made with the Canadian Geological Survey for one of their men to accompany us I suppose it would be on some such basis as your paying the man's expenses and the C.G.S. his time. The C.G.S. to be put in possession of the results of the survey while you would reserve rights to mineral discoveries.

I do not care for the idea of any intruders coming even if Dr. Sandberg was along, without him and considering my ignorance of the subject I can see all kinds of complications that might arise.

The second alternative of going ahead and preparing things generally is the one that strikes me most favourably at present. You are the one to decide on this and though objections on the wasted time and late return would be justified, yet the time would not be so badly wasted in that knowledge and experience gained first hand on the ground would be valuable.

The third alternative of losing a year I do not like to entertain. A year is a year in that region but little can be done in one season at best and the first year might be unprofitably spent anyway. I wrote to Lion to get his views on the subject and delayed writing until I heard from him.

In the first alternative he could not express any opinion but realized the complications that might arise.

The second, I put up to him that we should start out the first year on our own account, but he can't afford this, nor I for both of us and he agreed that it would be rather too much to expect from you.

Dr. Sandberg says he will know by the New Year how he stands. I actually hope he will be in the clear and these alternatives will not be faced.

I should be glad to know what think about it in any event.

With best regards to all,
Always yours affecty,
George M. Douglas

Douglas, Ariz., Dec 20<sup>th</sup>, 1910

Dear Cousin James,

Your letter of Dec 14<sup>th</sup> just arrived and I was much interested to hear about your meeting with Mr. Camsell of the Canadian Geological Survey. You did not say whether you discussed the question of this exploration or whether they would be willing or care about sending a man along with us.

As regards the question of wintering, whether at old Ft. Confidence or at Fort Norman, I was glad to have some experienced criticism on the subject.

It was my original scheme to establish some kind of winter quarters of our own on Great Bear Lake, but after careful consideration of the question, came to the conclusion that Fort Norman would be better. As Camsell says, and I realized, it would mean two extra men to pay and provide for and the transport of a lot of stuff up the Great Bear R. and across the lake. It would be an expensive business and it is not certain that we should gain anything by doing so.

Our season's work would be terminated by snow on the ground rather than by ice on Great Bear Lake and we should probably be at the Dease River in any case before the freeze up so we might just as well go on the rest of the way to Fort Norman.

I have been thinking a good deal about a route to Bathurst Inlet via Great Slave Lake and the lakes to the NE — Clinton Colden, Lake Aylmer etc. I have been reading Back's Journal and at present I am still of the opinion that the route via Great Bear and the Coppermine R. will suit our purpose best.

Supposing then for the present that we start from some base on the Great Bear side, the journey is a sleighing proposition in any case; sleigh outwards and canoe back again and the time of our start in the spring would be governed by the time we wanted to arrive at Bathurst Inlet. This would be about the beginning of June so if we wintered at Fort Norman we would have to start a week or so earlier than if we wintered at old Ft. Confidence.

However, I should like to discuss the question of our first winter quarters with some of the H.B. Co men before coming to any decision and as for the Bathurst Inlet exploration I would not come to any definite conclusions until after the first summer.

Certainly I appreciate what Mr. Camsell says of the difficulty of exploring it and unquestionably the best way would be by sea; to fit up just such a small sloop as the Gjoa in which Capt. Rould Amundsen went through these very waters. By this means both Bathurst Inlet and Victoria Land could be explored: copper is reported in the latter place as well.
The Gjoa was an old fishing smack (?) fitted with a small auxiliary gasoline engine ( I believe she is still at San Francisco) her crew, 7 men all told, and they accomplished what all those costly expeditions of the earlier part of last century failed to do . . . the Northwest Passage. Of course, Amundsen had the benefit of their experience and knew how to profit by it – just the same as we have the benefit of Amundsen's.

However, we are drifting rather far away from the simple original proposition of a trip by Sandberg and myself.

In any case we must come to some decision now. I shall see Dr. Sandberg this week if possible and find out how he is fixed.

I know a man out here who might be a good man for geologist, Tovote by name. I met him at Morenci first and he was working on the geological staff on some extra work, then went timbering in the mine for lack of a better job and in expectation of more geological work. Finally he got a job at Polaris in the engineering dept., but it seems that he disagreed with the mine engineer and in consequence only lasted a few days. I once asked Juelte (??) what he thought of him? "Plumb crazy', he said. But a man is often thus defined for no better reason than a difference of opinion. *Quisnaus igitur sanus?* Until we have some absolute standard for comparison I would consider Tovote less crazy than Smith. And a man needs to be crazy in the opinion of a certain class of people before he is fit to start out on such a trip.

Charlie Mills said he was a clever geologist but unbalanced. Charlie is an exceptional judge of men. So far as I can tell Tovote carries this general dislike by the tactless unwillingness to admit that American methods in everything are the best on this earth.

I enclose an article by him which may give you some idea of his geological knowledge. I am no judge.

He is a German educated at some university there, about 32, married with a child.

At present he is dead broke and working in the mine at Bisbee.

I asked him how he would care to go on such a trip (I did not mention your name in connection). He said he would be glad of the chance if he could get enough money to send his wife to Germany.

I do not know the man personally very well. I believe him to be pretty energetic and I believe we would get on all right.

Let me know what you think of this. I regret that I left the book you mention at Northcote Farm and I fear inaccessible but shall write to Mrs. Allen to ask her if she can locate it.

I am glad you saw Sir T. D'Shaughnessy. Lion's ship will be at Vancouver from Jan 7th to Jan 25th and again Mar 30th. He should leave the latter date.

I shall write as soon as I see Dr. Sandberg but since our last interview I am not very hopeful that he will be able to get away.

I shall be glad to hear what you think about Tovote.

With best regards to all and wishes for a Merry Xmas and Happy New Year.

Always yours affectionately,
George M. Douglas.

Douglas, Ariz., Dec 21st, 1910

Dear Cousin James,

Since writing yesterday I have received your letter of Dec 16th enclosing a copy of a letter from Mr. Camsell.

This is very interesting and certainly his advice about winter quarters should be given serious attention. This point of getting into touch with the Eskimos is very important and might be done with quarters certainly from Old Fort Confidence. Without this assistance our trip to Bathurst Inlet would be very difficult.

However we shall have a chance of discussing this question thoroughly before we come to a decision.

I received your book also. Thank you very much for sending it. I have only glanced at it so far, but from the style it promises to be interesting and intriguing.

I am going to Morenci tomorrow and shall try to meet Dr. Sandberg either there or Gumpas.

I have been thinking a good deal about Tovote since speaking to him yesterday. I have had him in mind for some time as a matter of fact but did not sound him until yesterday.

I believe he would prove a good man, he is enthusiastic and energetic and a pretty husky fellow.

Best regards and good wishes to all
Always your affectionately
George M. Douglas

Nacarozi, Sonora,
Dec 26th, 1910

My dear Lion,

Your ship should be due about the time it takes a letter to get to Vancouver so I shall let you know how things are progressing.

First, I have practically made up my mind that we shall start out this year. If something of that sort is going to be done the sooner one gets after it, the better: the first year will give us experience in any case, and that is mighty valuable.

Cousin James was in Montreal a short time ago and met Sir. T. O'Shaughnessy who said that he didn't think there would be any difficulty about your leave. I expect you will straighten things out before you sail in the end of January.

James Douglas also went to Ottawa and met some of the Geological Survey men there and gave them my plans to criticize. They seemed to think they were all right but advised us to establish winter quarters of our own on Great Bear Lake rather than return to Ft. Norman. This was my original intention but I thought Ft. Norman would be better for several reasons and I gave the subject serious consideration. However we shall discuss this subject thoroughly with the C.G.S. men and H.B.Co. men also before coming to any decision. Camsell, one of the Survey men who has been on Great Bear Lake, thought we might repair Ft. Confidence and winter there.

I think this was built originally by Simpson and Dease about 1830. I have no account of their trip. Sir John Richardson repaired the place and wintered there about 10 years later and I have his book.

Hanbury, the only man that I am aware of who has been up the Coppermine since, gives a picture of the places in his book and says only that "the chimneys are in good condition"!

This was five years ago and the H.B. Co. use it as a kind of summer hunting and fishing post: the game is plentiful there I believe.

With regard to any co-operation by the Canadian Geological Survey, Cousin James said only that Camsell knew the country very well – country generally, I mean – and thought that he would like to come with us had he not lately married and settled down in Ottawa.

Dr. Sandberg is still uncertain whether he can come or not. I have had another man in my mind for some time and about a week ago I finally sounded him to see whether he would care to come.

This man is called Tovote, he is a German and a pretty clever geologist. I have known him for several years, meeting him at Morenci first, where he was working on the geological staff of the mine there. He is a clever geologist but somewhat flighty perhaps. I really don't know him very well and never saw any such tendency in him. He is not popular with the men around generally mostly on account of a rather anti-American attitude. And it is a fact that these Americans in the west are inclined to look down on German educated men. Certainly there is no reason why they should unless a consciousness of the German superiority.

However, that doesn't worry me any with Tovote. I never saw much of him at Morenci because I knew all the men there that I wanted to know anyway.

I believe Tovote would be a pretty good man for us. He is about 35, married with no child [in his letter December 20, Douglas states Tovote has a child], working in the mine at Bisbee just now, dead broke, and says he would be glad of the chance to come with us.

I have put it up to Cousin James if he wants to stand for the expense in connection with this man. I shall include him. He is a pretty husky fellow and I never saw that he was afraid of exerting himself.

So I think it likely we shall take him whether Dr. Sandberg can come or not. I shall probably leave here about the end of January. I shall get in touch with the H.B. Co. authorities and make a trip to Winnipeg to fix up for the transportation of our stuff and get all the advice I can on equipment, provisions, etc., what we can take and what we can get on the spot at various trading posts.

You had better fix it up to leave the ship as soon as you come back in March and join me at Northcote Farm. It will not be the best time of the year there but we shall put in a few weeks pleasantly enough and prepare ourselves a little for what we may expect. We should be on the ground ready with all our stuff at Athabasca Landing by the end of May.

[two pages on Mexico left out here]

Dr. Sandberg is still doubtful. I constantly hope he can manage to get away.

Well, I trust we shall meet in Peterboro in a few months. My address is Douglas, Arizona, that is my headquarters. I shall write again before you sail.

Best wishes for the New Year,
Always your loving brother,

George M. Douglas.
R.M.S. Empress of Japan,
En Route, Kobe to Nagasaki
February 13th, 1911

My dear George,

We arrived at Yokohama on Feb 11th after a very rough passage across the Pacific. I had an interview with the Manager concerning my leave, which he has granted me – 12 months leave with a probable extension of 6 months – Sir Thomas Shaughnessy had written to him about me so he could not well refuse, but spoke about "unusual length of leave" etc but eventually granted me the leave, so I hope to leave this ship on April 4th and will come right through to Lakefield. This is very satisfactory, and now for the Wilds of Northern Canada, it seems so hard to realize that this trip is really going to take place, it is owing to being so tied up to a ship tho' – one gets into a rut on a run like this & you know practically what to expect for months ahead, so a change (and this will be one!) does one good.

I am sending this letter back by the "Empress of India" and I hope it will find you at Lakefield about the middle of March; how strange it will be to meet you there after all these years. I remember so well our departure from there in April '94 – two months from now and I hope we shall be together getting things together for the trip North. Lovely weather here (Inland Sea Japan) and it is a pleasant change after the Stormy N. Pacific. With love

Your loving brother,
Lion


[On April 9th, 1911 George met Lionel in Peterborough, Ontario, the closest train station to Northcote. When the train came in and various passengers came towards him, George didn't see Lion, but he noticed one rather distinguished looking man with a dark pointed beard, and before resuming his search, wondered what this foreign–looking man was doing in Peterborough. Douglas had resumed his search, when suddenly his shoulder was seized from behind and he turned to see that the "foreigner" was his brother, whom he hadn't seen in thirteen years.

In his earlier letters Douglas thought that Sandberg might be unable to accompany them, so was considering another man, Tovote. Sometime in January or February, as the following letter documents, Sandberg decided to go on the trip. I was unable to ascertain exactly how he cleared up his affairs and whether or not James Douglas had any role in this. I suspect he did, but it is unproven.]

Gumpas, Feb 20th, 11

Friend George,

I have given my resignation to Mr. Beauchamps who is just back from his trip east. I have also suggested Tovote as my successor.

And now I wish I was out of this place and on our trip. If anything happens to this concern as you say I will start for the north right away. It is not likely that anything will happen though. The only interest I have in staying here now is the money. It is a strange concern this, and the people that have been out are putting up the money deserve a better treatment than they have been getting. And now I do not believe that they will find a mine here to reimburse them for their outlay.

But this matter is behind me now. And our trip is what requires our attention.

I got a letter from James Douglas in which he also suggests that I visit the Lake Superior copper mines, which of course I shall be more than glad to do. And the data I can gather there will be of great importance, and I can get an idea of the copper occurrence there better by observation than through the literature.

I leave here the 1st of April, and I think that I can go through the Lake Superior District and meet you at Winnipeg or Edmonton about the 1st of May. If you think it necessary I can come earlier, and will have to cut out New York. There is some things I wanted to buy there, but I can get them also no doubt in Chicago.

I imagine it will be a rest to get into open country, plains, lakes & forests and not see everyday red mountains, while one is penned up in arroyos most of the time.

No news from the east of war, if one exists. Those people are not capable of great efforts.

Yours sincerely,<br>Aug. Sandberg

Edmonton  May 22<sup>nd</sup>, 1911

Dear Cousin James,

Our outfitting is finally complete and our freight already on its way to Athabasca Landing. We follow tomorrow on the stage.

After talking matters over with the H.B. Co and being assured by them that there will be no difficulty getting Indians at Ft. Norman to help us up the Gt Bear R. and across Gt Bear Lake, we have decided to winter at the Dease R. if we possibly can.

It is not certain whether we could get a boat to suit us on the way down and I think the transport of our stuff from Ft Norman can be best done by canoes in any case. So, I got another big canoe here. I think we can convert the crates in which the other canoes are packed into fairly good boats by covering them with waterproof canvas which I am taking for that purpose. I did not think of this at the time I had the canoes packed but I had exceptionally good crates made that the boats might be well protected on their long journey.

So our plans will be to engage 5 Indians at Ft Norman and pickup a few more to help track up the river.

There will be 2 Indians to each of the "crates" and Indian and Dr. S. to the big canoe while Lion and I can each take one of the others. On our arrival at the Dease R. we shall send back all the Indians but two, who will remain with Lion until they build some kind of a house for winter quarters. As soon as this is done their services will be dispensed with.

Dr. Sandberg and I shall go ahead to the Coppermine R. as soon as we can and with the assurance of winter quarters to fall back on we shall stay out till about the end of Sept.

We can start early the following summer and make investigations along the line indicated as most profitable by the previous season's work.

    We should leave about the end of August and could probably get as far as Ft Simpson before the river froze. The H.B. men advise us to make that point if we can rather [than] stopping at Ft Norman.

Our outfit as completed is as follows.

Canoes

| | Probable capacity | as we shall load them |
|---|---|---|
| 1 19' x 42" x 18"  basswood canoe | 1600 | 1600 |
| 2 18' x 36" x 15" cedar and butternut canoes | 550 | 1100 |
| 2 12.5' x 40" x 16" canoes | 1000 | 2000 |
| | | 4700 |

I intended to take a light 32.30 rifle and a shotgun but have left these behind. Our armament as it probably will be leaving Athabasca Landing. We also left a .32 and a .38 revolver.

3 x 8mm Mauser rifles, 2 x 7mm Mauser rifles, 1 x .22 rifle, 2 x .22 Stevens pistols with skeleton stock

2500 rounds 8mm (200 lbs), 2000 rounds 7mm (140 lbs), 5000 rounds .22 (35 lbs)

Instruments

2 Brunton Transits, 1 Sextant, 1 artificial horizon, 1 spirit boat compass, 2 boat compasses on gimbels, 3 pocket compasses, 1 canvas bag [?], 1 aneroid barometer, 2 pedometers, 1 Locke hand level, 3 min. recording thermometers, 3 ordinary mercury thermometers, 2 boiling point thermometers, 1 clinometer, magnifying glasses

Food Supply

Flour – 700 lbs
Self rising flour – 100 lbs

Desiccated potatoes – 100 lbs
Desiccated vegetables – 20 lbs
Soup tablets 10 lbs

Bacon – 400 lbs
Pork – 200 bs
Sugar – 500 lbs
Rolled Oats – 160 lbs
Beans – 240 lbs
Lard – 310 lbs
Corn meal – 100 lbs
Hominy – 110 lbs
Baking powder – 65 lbs
Tobacco – 95 lbs
Condensed milk – 50 lbs
Dried apricots – 30 lbs
Coffee – 30 lbs
Tea – 45 lbs
Raisins – 110 lbs

Chocolate biscuits – 25 lbs
Desiccated raspberries – 20 lbs
Horlicks malted milk – 3 lbs
1 doz bottle brandy
6 bottles rum
Dried apples – 120 lbs
Dessicated onions – 20 lbs
Maple syrup – 14 lbs
Peanut butter – 3 lbs
Sea biscuits – 200 lbs
Milk chocolate – 100 lbs
Bakers chocolate – 60 lbs
Cocoa – 50 lbs
Erbwurst – 110 lbs

<u>Various</u>

20 gallons coal oil, 80 lbs candles, 55 lbs Sunlight soap, 35 lbs matches, 3 dozen paddles, 50 yards canvas for roofing, 40 yards canvas for canoe covers.

All above weights are gross weights as packed. Total weight for freight bills 4822 lbs

<u>Outfit of clothes for each man</u>

1 Mackinaw coat, 2 extra heavy wool pants, 2 pair Duxbak pants, 1 Duxbak coat, 1 leather vest, wool lined, 2 heavy sweaters, 1 medium sweater, 3 suits heavy wool underwear, 3 suits medium wool underwear, 3 suits cotton underwear, 18 pair heavy wool socks, 6 pair light wool socks, 4 heavy wool shorts, 1 waterproof shirt, 2 pair boots, 1 pair long Larrigans, 1 pair short Larrigans, 1 sleeping bag, 1 light "HB" mattress, 1 double blanket, 1 single blanket

Total weight for 3 including some ammunition, books, etc mixed with clothing

in bags 764 lbs
Tool Supply

2 One man saws, 1 Bucksaw, 2 Crosscut saws, 1 Rip saw, 2 x 3 1/2 lb axes, 2 x 3 lb axes, 5 x 1 3/4 lb axes, 1 Hatchet, 1 Brace, 1 Drawknife, Various bits, 3 doz files, 2 doz files (for trade), 1 Hacksaw, Various chisels, Various palms, shoemakers tools, lasts, etc, 2 x 4 1/2" nets, rope, sinkers, floats, fishing line, hooks etc etc in a large supply
I got a set of cooking things at N.Y. and have added a duplicate set here (only much inferior).

+ see Editor's note on p. 357

Our total outfit at present weighs 6611 lbs but we shall reduce this a little at Athabasca. Also we shall leave a certain amount at Ft. Norman so that our weight leaving there will be about 4500 lbs more or less.

The cost of the expedition to date as follows.
(figures in columns to right are what expenses I have paid)

| | | Total Expenses | |
|---|---|---|---|
| Canoes and canoe fittings, spars, spare paddles, crates, etc | | | |
| Lakefield boats | | 240.25 | 240.25 |
| Edmonton | | 136.75 | 136.75 |
| Food Supply | | 755.30 | 279.10 |
| Clothing, Blankets, Boots etc | | 306.20 | 159.48 |
| Various camping equipment, tents, canvas covers, waterproof pack sacks and waterproof canvas bags, fishing nets, etc etc | | 330.35 | 330.35 |
| Rifles and ammunition etc | about | 500.00 | 170.15 |
| Customs | | 187.60 | 187.60 |
| Freight | | 326.04 | 29.95 |

|                                                           |         | 326.00    |
|-----------------------------------------------------------|---------|-----------|
|                                                           | ———     |           |
| RR fares Including trip to Ottawa                         |         |           |
| and Montreal (r.r. fares only)                            |         |           |
| Trips to N.Y. Dr. S. trip to L. Superior                  |         |           |
| to Edmonton                                               | 406.75  | 406.75    |
| Hotel Expenses                                            | 182.00  | 182.00    |
| Photo Supplies                                            | 88.05   | 88.05     |
| Medical Outfit                                            | 17.25   | 17.25     |
| Tools, Coal Oil, Compasses,                               |         |           |
| various instruments                                       | 119.40  | 119.40    |
| Telegrams                                                 | 8.65    | 8.65      |
| Advance by H.B. Co                                        | $1000   | $2355.75  |
| Received by cheque from you                               | $1500   | 326       |
|                                                           |         | 2681.75   |
| Balance in hand                                           |         | 144.25    |
|                                                           | $2500   | $2500.00  |
|                                                           |         | 2681.75   |
|                                                           |         | 2500      |
|                                                           |         | 181.75    |

Amount paid by you for various N.Y. visits must be nearly $800.00. I have no note here of the exact figures.

So the bill to date is as follows:

|                                                           |           |
|-----------------------------------------------------------|-----------|
| Advanced by cash                                          | $2500.00  |
| N.Y. purchases                                            | $800.00   |
| H.B account, supplies, freight from                       |           |
| Lakefield here and from here to Athabasca Landing         | 941.62    |
|                                                           | $4241.62  |
| Cash in hand                                              | 144.25    |
| Cost to date.                                             | $4097.37  |

Letter of credit $5500.00 so I have still to my credit $144.25 and $3558.38

credit.

Further expenses

| | |
|---|---:|
| Edmonton to Athabasca | $24. |
| Athabasca to Ft Norman | $261.00 |
| Freight Athabasca to Ft Norman | $774.00 |
| Canoes | $174.00 |
| | $1233.00 |

I do not know what the Indians will cost yet but I shall pay them partly in goods we have already bought.

I note I have forgotten to include photographic outfit as follows.
1 4 1/4" x 6 1/2" Kodak with Goetz lens and Sector shutter
1 3 1/4" x 5 1/4" Kodak with B&L Plantograph
A spare R.R. lens and shutter for each camera
7 doz rolls of 4 1/4" x 6 1/2" film each packed in tin waterproof case
2 1/2 doz rolls 3 1/4" x 5 1/4" same packing
1 5" Developing Tank with spare tank and spare reel
20 lbs Hypo
1 1/2 doz packets developing powder for tank

Total weight of above outfit about 120 lbs.

On our arrival here the H.B. Co. district manager had told us that the boats would leave Athabasca Landing on June 7th.

He went down to Athabasca Landing himself a few days ago and we got a wire from him to say there were no June 7th boats and to come to the landing at once, he had a canoe and man waiting for us to take us down the river to Grand Rapids where he thought we might overtake the first brigade of scows that had just left.

His assistant here wired that this was impossible for us as we had all our freight still at Edmonton. The manager, Fugh by name, wired in reply to send freight as soon as possible and to leave Edmonton ourselves on Tuesday.

Everything was ready then and it was sent out at once and we go tomorrow. We should arrive at Athabasca Landing on Wednesday evening and our freight should be there by that time.

I do not know for certain what arrangements have been made nor the reason for cutting out the June 7th boat.

The asst. manager, a Mr. Max Hamilton, thinks we can send part of our freight by a brigade of bateaux leaving on Thursday and take the rest in the waiting canoe and in our own big canoe. The other canoes have been sent ahead already.

He seems to think that we shall likely overtake the first brigade at Grand Rapids and transfer freight etc to them. Or if not at Grand Rapids certainly at Smith's Landing.

I had got the big canoe before this mix up happened and you can imagine that it is a valuable thing for us to have now.

Robt Service, the author, is coming with us and has suffered like ourselves from the change of orders. I met him today he is rather a pleasant sort of fellow.

We leave early tomorrow. I shall write a note from Athabasca Landing. Good bye for the present and best regards and love to all.

Always yours affectionately,
George M. Douglas

[There are several items that Douglas left off the above lists. Receipts for these items exist. I include the items here for the sake of completeness.

The following were purchased from J.J Turner and Sons in Jan–Feb, 1911. Apparently, the toboggans and snowshoes were left behind as Douglas states (*Lands Forlorn,* p. 9) that they took nothing in the way of toboggans, snowshoes, or fur clothing. Douglas was rather upset that they left behind these high–quality items.

1 pr Snowshoes #1, 1 pr Snowshoes #9, 2 x 7 ft. Toboggan, 2 Boat rugs, 1 Wool rug, 1 pr Grey blankets, 1 pr grey blankets , 1 pair 7ft skis, 25 yards 44" Duck, 1 Canoe tent 6 x 7 1/2 – 8 oz Duck complete with poles and pegs,

1 Canoe tent 6 x 7 1/2 – Waterproof silk complete with poles and pegs, 1 Shelter tent 9 1/4 x 7 – complete with poles and pegs, 3 Mosquito Fronts. A receipt dated May 22nd, 1911 from the Hudson's Bay Co., in Edmonton lists the following items of which much of the food is included in Douglas' list above. 2 sweaters, 37 prs socks, 4 suits underwear, 18 shirts, 15 yds canvas, 3 prs towels, 4 1/2 yards toweling, boots, 11 pair underwear, 10 yards cheesecloth, cloth, 3 bot. brandy, 6 bot rum, 2 bot. lime juice, 30 lbs flour, 30 lbs sugar, 24 lbs rolled oats, 40 lbs bacon, baking powder, lard, 3 lbs coffee, matches, pepper, salt, tea, cream, 700 lbs flour, 1 c.s flour, 600 lbs bacon, 500 lbs sugar, 4 sks rolled oats, 2 c/s honey, cornmeal, 5 lbs rice, 240 lbs beans, 3 c/s lard, 40 lbs. H.B. powder, 4 tins syrup, 20 lbs. coffee, 25 lbs tea, 1 pipe, 1 c/s cream, 18 sacks, 20 duck sacks, 100 lbs evap. apples, 1 c/s apricots, 100 lbs raisins. 1 doz yeast, 1 case sunlight soap, 15 case matches, 2 cases candles, 20 lbs salt, 3 jars fish paste, 6 jars peanut butter, 20 evap. onions, 75 imperial tobacco, pipes, 20 imperial mixture, a liquor permit ($3.00), 2 prs., 4 point blankets, 3 Birk G. Roll ups, 2 G.M. cots, 4 c/s 16 x 13 1/2 x 7 1/2, 10 yards Black Watertight and a mysterious doctor's fee of $1.50.

Michie and Co. of Toronto provided several food items carefully packed in 5 and 10 lb biscuit tins which were soldered shut. The tins were then arranged into 24 packages ranging in weight from 43 to 60 lbs (total weight 1167 lbs). These items, as Douglas stated, withstood shipping better than other packed foodstuffs. They were all shipped directly to Edmonton by Canadian Pacific Rail.

A short packing list of items to leave Fort Smith yields the following additional goods: extra axe handles, a shovel, one or more Reflex baking ovens, tarpaulins, notebooks.

Additionally, he brought along a small scale as he gives fish weights to 1/2lb. It hangs from the shelf in the upper left–hand corner of the photograph on page 390. A receipt for items bought at Fort Norman exists but it is likely that many, if not all, of those things were for the Indian trackers and their families.]

99 John Street
May 23rd, 1911

Geo. M. Douglas, Esq.,
    C/o District Manager,
        Hudson Bay Co.,
            Edmonton, Aberta.

My dear George:

I was glad to get yours of the 13th, as I was wondering what had become of you. I have nothing to say beyond wishing you all Godspeed and a happy return, whether with or without copper specimens matters little. Your friend Grier will not part with his original impression of your likeness, but he is going to make me a copy. I can trust to his judgment as to whether it will be a real reflection of the original.

My best regards to your companions, including the fourth member of your family, who I hope you will (when found) find to be a congenial companion.

Yours most truly,
James Douglas

Fort MacMurray, June 7th, 1911

Dear Cousin James:

We have finished the first stage of our journey and I shall get this ready as there may be a chance to send letters out in a few days. We got all the more important part of our outfit loaded on two big canoes, our own "Aldebaran", and a similar big one left for us by the Hudson Bay Co. We took about 1000 lbs. on each canoe. Lion and I took the "Aldebaran" while a half–breed, Pat Prudhomme by name, took the other. Dr. Sandberg and Robert Service took turns helping Pat, and as passenger on the "Aldebaran".

Our hurry and worry reached its culminating point, segregating and repacking stuff to suit the altered plan and loading it from a muddy bank.

It had been a stormy day with high wind and frequent heavy showers but came off a beautifully fine evening. We pushed off at 6.30 P.M., and the contrast was certainly very pleasant; the lovely evening and the sudden "cut-off" of worry; we had cast-off and no use worrying about unalterable things. The Athabasca River for the next 100 miles is fine and smooth, high banks, heavily wooded with spruce and jack pine, poplar, and balme-of-gilead. We had a very pleasant trip down the river and got to Pelican Portage, at the beginning of the rapids, on the evening of the second day. We were still about 40 miles from Grand Rapids, but Pat suddenly got nervous and wanted some one to take us through the Rapids, so he landed here and went to try and find a man while we investigated a gas well in the neighborhood. The Government were making some borings for oil about 16 years ago and they struck gas. Some one set it on fire and it has been burning ever since, a flame about 30 feet high through a 2" opening. For they have made some haphazard attempts at capping it and have reduced their 6" casing to 2", so far a great waste. Pat could not find a man, so we started off in some nervousness, Lion and I leading in the "Aldebaran". But our anxiety was quite unnecessary, the rapids did not amount to much, and we went through all right. It was swift water all the way to Grand Rapids now, and we arrived there at noon, the third day from the Landing, about 140 miles. The "brigade" was still there, they had only a couple more scows to lower down, The Grand Rapids are certainly worthy of the name; an island about half a mile long makes two channels. The west channel is the main river, and no craft could run that swirl. The east channel is not so bad, and by discharging all their cargo on the upper end of the island, the scows can run that empty. The cargo is taken to the lower end on a rough tramway.

The "brigades" consisted of about 20 scows; these scows are about 50 feet long and 12 feet wide, about 3 1/2 feet deep. They are manned by five men as a general thing, four men at sweeps and one steersman. Going down bad rapids they put two men on the steering sweep – an immense thing sometimes, over 40 feet long, and a couple of men at the bow. We found our two canoes, the "Polaris" and "Procyon" here all right but somewhat scratched up. I can understand now why they did not want to ship them in crates; but I wish they had told me about it before; certainly they had plenty of time.

We finally left Grand Rapids on May 30th. Lion, Sandberg, Service and I were together on a scow smaller than the others. Besides ourselves there were eight or nine more passengers, some Northwest Mounted Police, a

Athabasca Brigade

missionary, and his family, and a farm instructor; they were divided up on the other scows.

It took us just a week to make the next ninety miles to MacMurray. There are many bad rapids and at some of these the scows had to be lightened. It was certainly a most picturesque and interesting life. I think without exception the most interesting piece of travel I ever had. The day's routine would be to start about 7 after breakfast, tie up for lunch, and again at about 6 P.M. for the night. We made our beds ashore under a mosquito net. There was a scow especially for cooking and the meals; the latter were not bad considering the circumstances. We put the "Polaris" on board one scow and towed the "Aldebaran". Lion and I ran all the rapids in the "Procyon"; it was quite fine. The first of the big rapids after Grand Rapids, Lion and Sandberg went down in the "Polaris" while I took Service in the "Procyon". We went down in fine style but both Sandberg and Service declined to run any more rapids unless they had to, so Lion and I went by ourselves in the "Procyon" thereafter. We saw moose several times; as usual the only time we could have shot one we did not have a rifle with us. The Indians got one moose, and we saw a good many bear also. The banks are high, usually terraced. At Grand Rapids they are sandstone, with other "nodules" of harder sandstone imbedded, quite curious [these are called concretions]. These tar sands are also very interesting there is less tar in the mixture than I expected to find.

I was quite sorry when our journey came to an end at Fort MacMurray. The steamer "Grahame", that will take us to Smith's landing on the Slave River, was tied up waiting for us, and we moved our gear on board her at once. She is a flat bottomed stern wheeler, 140 feet long, the regular type seen on western rivers. Lion and I have a cabin together and Sandberg and Service, they are not very big. We have our meals aboard. It is quite comfortable but less interesting than our scow life. We shall leave here in a couple of days. Shot's brigade that has the rest of our supplies, is still at Grand Rapids, and will not overtake us here as we hoped. His scows are taken in tow by a little missionary steamer, and I hope at least that they will get to Smith's landing in time to catch the steamer at Fort Smith. All our food supplies are with Shot except bisquit, chocolate, erbwurst, dessicated potato, etc, we brought about a hundred pounds of each of flour, sugar and beans, about 40 lbs. oatmeal; no bacon, except about 20 lbs; no dried fruit of any kind.

Fort MacMurray has a beautiful situation on the rather elevated flat point between the Athabasca and Clearwater Rivers. The steamer and scows are tied up to a big island in the river opposite the post. The post consists of a few log houses, a small Hudson Bay store, and another small store run by a "free trader", a Miss Gordon. There are a number of Indian tents and teepees, and the whole is very interesting, with plenty of "character". The dogs are quite a feature of these camps; big, wiry brutes, surly and suspicious, but nevertheless amenable to kindness. They are apparently very destructive, and the Indians make a kind of tripod of saplings and hang anything that might be chewed up out of the dogs' reach. The interior economy is much like some Mexican camp; same smouldering fire and same, or very similar, old women squatting in front of it, making bread and smoking. Only she smokes a pipe instead of cigarettes, and the bread is a kind of bannock instead of the thin "tortillas".

The woods around here are mostly poplar, which attains a greater height than I have seen before anywhere. Birch is the only hard wood of any kind and there is not a great deal of that.

The season has advanced very quickly; the trees are in almost full leaf. But there is still ice piled up along the shores in places, usually below rapids, where it is sometimes ten or twelve feet high. We have seen ice at intervals ever since leaving Sudbury in Ontario. The days are getting very long, even now there is hardly any night at all. At ten P.M. one can read and write without any trouble.

I shall let you know again from Fort Norman how things go with us.

> Love to Aunt Naomi and to all.
> Always yours affectionately,
> George M. Douglas

> Fort Smith, June 23rd, 1911
> On Board S.S. "Mackenzie River"

Dear Cousin James:

We are now at the head of navigation with a clear run to Fort Norman. We left MacMurray on June 9th, and had a pleasant journey in the "Grahame" to Smith's Landing, 16 miles to the south of this place. The Athabasca River is not particularly interesting below Ft. MacMurray; the shores become lower and are heavily wooded with white spruce, poplar and cottonwood. The current is fairly swift and there are many islands and sand bars. It was a pleasant change to get on Athabasca Lake and we spent two days at Fort Chipewyan, one of the oldest and most important of the Hudson Bay Co.'s posts. It is situated on a rocky point on the northern side of the lake, granite or gneiss, and high rocky islands off it; something like the Thousand Islands, but rather higher and the trees look different. The post consists of about thirty log houses, beginning at one end with the Roman Catholic School and mission, and stretching along the shore, little sandy beaches alternating with rocky points, and the Hudson Bay establishment on a high rocky point at the end. It probably looked much as it does now when Franklin came here in 1819. He arrived at the end of March; it was stormy weather, and he had to take shelter behind one of the islands until the weather moderated enough to let him cross the lake. It is only a few miles across, but it must indeed be a bleak place in winter time. It was the W.W. Co.'s post then and probably there were more Indians in those days than there are now. Whether they have improved since that time I should think very doubtful. Franklin's account of the dirt, indolence, and lack of thrift reads like a description of those we have seen so far. I talked with the fathers at the Roman Catholic mission. It is indeed a thankless sort of work they have undertaken. The school is quite large, probably there are about 80 boys, and as many girls. The brothers were mostly French; there was one Irishman, an amusing old gossip. It was like a page out of Parman, or your own books, describing the early missions. I must say that the children looked promising; the good fathers have taught them manners at least. I don't know whether the impression will be lasting; they go back to their woods and teepees and at the end of a few years are apparently so little the better for it that one wonders what is the good of it all. I do not know how long the mission has been established, nor what the Indians were like before they came. Judging from the early accounts, much as they are now – nor what they will be like after a generation or two. But there can be no question of the self–sacrificing courage of the fathers. The reward lies in the doing; it strikes a casual observer as a most thankless task. There was a little English chapel there, and an ineffective parson; it seemed for the benefit of the white people, Hudson Bay clerks, and "free–traders", etc.

We ran down the Rocher River to the Peace River. The Slave henceforth is a great river, much larger than I expected, and comparable with the St. Lawrence above Montreal. We came to the limit of the

"Grahame's" run at Smith's Landing; all the freight is landed here and taken by wagons to Fort Smith below the rapids. Smith's Landing is a very pretty place. The river makes a great bend and widens out to more than a mile; then narrows again to the Cassette Rapids. These are at the lower end of some islands; a little above these islands there is another row of smaller islands strung across the stream, and small rapids between them also. Lion and I ran these in the "Procyon" and landed just above the main rapids. A fine sight they are, but a most hopeless place to get into. Next day we took the Doctor down to look at them and on our way back, coming up the small rapid we upset. Lion and I had come up them all right the day before, but a sudden lurch on the Doctor's part brought her side under at the critical moment, though I should hate to tell the Doctor it was his fault; it was really my own for letting an inexperienced man paddle at all, in a place like that. We went under anyway. I tried to empty the water out, but we were in a bad eddy and I was hampered by boots and a good deal of heavy stuff in my pockets, and could not manage it. We had no control over her in that current and as we were. Fortunately, we were not far from shore, so I swam ashore, and took my clothes off. Lion staying with the canoe, gallantly, in the meantime, and holding her in the eddy, and we soon ran her ashore then. The Doctor had swam ashore earlier; it was really all he could do. Not till we had emptied the water out of the canoe did I think of the camera which happened to be in her. Lion and I had both stuck to our paddles. I had saved the Doctor's coat and my own hat; Lion had rescued my sweater at some trouble to himself, but the camera had been quite forgotten. We paddled around as soon as we could hoping it might still be afloat, and rescued the Doctor's paddle just as it was going over the fall. But we never saw the camera again. It is a serious loss to us, as we are fixed, besides being a valuable camera, 4 1/4 x 6 1/2, and with a Goertz lens. It might have been even worse than that; it was an ugly place for an upset to occur, just above the main rapids; the eddy was our salvation; a humiliating business all around, after the rapids we had come through, to turn over in a place like that. Fortunately I have the 3A and a good number of films for it, but that 4A was our main standby.

On the scows the Hudson's Bay Co. certainly handled freight well. They had sufficient men and knew how to get the work out of them. It is just the reverse on these steamers; they are painfully slow, and generally careless. We had some thoughts of running the rapids to Fort Smith in the "Procyon" (Lion and I) but gave it up as being unnecessarily risky; the river is much cut up by islands and there are a number of portages; the channel is an intricate one. So we let the "Procyon" make the passage ingloriously at ease on a wagon, like her sisters "Polaris" and "Aldebaran". Sandberg and Service had already gone over. Lion and I saw our freight and canoes loaded, then walked over; a pretty but not particularly interesting road through thick woods all the way. The balance of our stuff, which had been brought down by Capt. Shot from Athabasca Landing to Fort MacMurray, and by the Roman Catholic mission steamer from MacMurray to Smith's Landing, overtook us here, so we have a chance for the first time to see all our stuff together and check it over. A good deal of it is in waterproof bags, clothes, etc., some provisions, rolled oats and beans, rice and cornmeal. Of these "kit" bags we have about forty. It was necessary to stowe a good deal of stuff that way, on our canoe trip from Athabasca Landing to Grand Rapids, and they got pretty badly trampled, and a few torn on their subsequent scow, steamer and wagon transport. Dried fruits and other supplies packed in boxes arrived very smashed up. It was packed by the Hudson's Bay Co,. and very badly packed at that. But the only actual shortage is a case of raisins, a serious dimunition of our fruit supply, nearly 50%! The stuff I got at Toronto came through best of all; it was most satisfactory in every respect; good quality, carefully packed, and cheaper than I could have got it at Edmonton, even including freight to that point. I regret now that I did not get all my supplies there. [these items came from Michie and Co., Ltd. of Toronto and were hermetically sealed by them in five and ten pound biscuit tins and then arranged into 24 packages of 40–60 pounds each. Letter April 7th, 1911; Invoices April 13th, 1911]. We have been learning lots of things on this trip. Our canoes are in good shape, especially so considering the trip they have had already, but Lion and I handled them ourselves at every possible point. We re–varnished the

"Polaris" and "Procyon", for defence rather than ornament, and painted the "Aldebaran" a dark red inside and dark green out, so our fleet looks very smart. The "Polaris" and "Procyon" have been the admiration of the river since we started – the first canoes that ever came down here. We met Mr. Brabaut, the Hudson Bay Co. manager for the Mackenzie River district. Fort Smith is his headquarters, and he is making the trip down in the steamer. We learn that two or three Englishmen, Hornby and Melville, made a trip into Gt. Bear Lake last summer, intending to winter at old Ft. Confidence, and to come out again this summer, hunting and trading seem to have been their object. I expect we shall see them at Fort Norman, and we shall learn what kind of place they put up for winter quarters, and whether they have any york boat or other equipment we could avail ourselves of. For it is certain that if we are going to winter at the Dease River we shall have to get a large boat. But if we are likely to lose any time taking our stuff up the river we shall come back to our original plan of wintering at Ft. Norman, as we can get in a longer time at the best time of the year by doing so; better, for instance, spend from Aug. 1st to Sept. 1st, rather than from Aug. 15th to Sept. 15th. However, we shall settle this at Fort Norman, when we shall find what means are available. We can scarcely use canoes; we have nearly three tons of stuff—none too much of anything. In fact I shall add to our stock of flour at Ft. Norman.

I hope this letter is legible and intelligible. The mosquitos are very bad here; and one has to write in a head net, and stop every word or two to brush them off one's hands. The fly dope we brought with us is fairly good for a few minutes but seems to hurt the skin if much used. I haven't tried your suggestion of turpentine and castor oil but it would probably be much like the stuff we have. Canvas gloves and head net are best when they are thick.

Fort Smith is not an interesting place. It is on the edge of a thickly wooded plain 130 feet above the river. The "Mackenzie River" is anchored alongside at the foot of a sandy hill, and just below the last rapids, the "Rapids of the Drowned". These are not much more than swift water but very dangerous from ledges and whirls. The "Mackenzie River" is a new boat, stern wheeler, 125 ft. long and not so comfortable as the old "Grahame" for passengers. The cabins are very small and stuffy. Lion and I are together in one about as big as a good sized soap box; the bunks are 5' 9 1/4" long and I am just 6 feet. I would camp on shore if we could, but we can't for various reasons, stowage of our stuff aboard, etc. The Captain is an amusing old timer of the North. I heard a noise in the galley as I passed the other day, and looked around in time to see him threatening the cook with a pistol! I don't know what the row was about; a feud between the captain and the cook seems the fashion with all the transports we have been on here. We have come across some amusing characters so far. There is a certain "Captain" Crean in charge of what is called the "New North West Exploration". It is run by the Government and its object is to explore the country extending northwest of the Peace River. The only exploring Crean seems to do is in the outfit of various incoming people for first chance at the liquor "permit". You are allowed only one gallon of liquor apiece; this allowance is known as the "Permit"; mighty good thing too, that they keep liquor out like that. Crean has a little launch that he runs up and down the river, tows scows for fur traders and affords everyone amusement by his accounts of his various exploits—a notorious liar and humbug. Also we have come across a certain Radford of the "Radford Arctic Exploration", backed according to him, by the Peary Arctic Club. He plans to go across to Chesterfield Inlet, then up somewhere to the east of Back's River and along the coast to the Mackenzie River. He has been fooling around this part for a year or more and the various Hudson Bay people have him summed up as an incapable idiot. A casual meeting with him inclines one to this opinion. [The Douglas party bought a small sheet metal stove from him, which was most useful in their cabin at Hodgson's Point. Harry Radford and his companion, George Street, left Fort Smith on June 27th and were killed in self–defense by Inuit at Bathurst Inlet the following year.] We hear of Hanbury who has made himself famous in the North as one of the best and hardest travelers that ever came into the country. He was a big and very powerful man, with great capacities for both eating and drinking. Like the old Norse heroes, he is held in honor for his exploits at both. He seems to have been a "booze-fighter" and maybe the trip to the North was to get away from it. I hear that he is now dead. [Hanbury died October 25th, 1910 of an apparent heart attack at age 45. According to the *San Francisco Examiner*, he had been in poor health for several months after he broke several ribs from falling on the liquor bottle from which he was drinking.]

Lion and I took a walk down the river about 9 miles today, pretty country but infested by mosquitos; we were well protected with head nets and gloves. Living as we do on the steamer they are not a serious trouble, but I can imagine what they would be like if we had to make a trip through the woods. Seton-Thompson's account of them, which sounds exaggerated, is at least sympathized with, if not actually concurred in; and on Franklin's trip Hood says it was an "evil of such magnitude that cold, famine, and every concomitant of an inhospitable climate must yield pre-eminence to it". We should leave here early in the week by June 28th at latest, and I hope we may get to Fort Norman by July 8th. I shall write again from there and let you know how we get on.

Love to Aunt Naomi and all,
Always yours affectionately,
George M. Douglas

[On June 27th, in Fort Smith, Douglas bought the following items: 1 block plane, 1 cake soap, 2 lbs chocolate, 3 rings for canoe, 5 lbs babiche, 2 dogsleds, 1 cut twine #9, 1 qt wood alcohol, 1 camp stove, 2 axe handles.

On June 30th in Resolution, Douglas bought: mooseskin, an axe, a large pan, a pipe, plus other misc items which are illegible on the receipt.

On July 2nd, in Fort Simpson, they bought cod line, rope, rigging for a boat, blocks, 3 pair moccasins, 1 moose skin, 1 young moose skin, tarpaulin, 2 window sashes, 6 panes of glass 7.5" x 8.5", some boards, 5 lbs of nails, 1 wool toque, 2 pails.]

The Hudson's Bay Company
Ft. Norman, July 7th, 1911

Dear Cousin James:

We left Fort Smith on June 28th and made good time the rest of the way to this place. The journey was a most pleasant and interesting one. I shall tell you about it some time, but at present I write in a good deal of a hurry.

We got a york boat, the same one that our steamer towed from Ft. Smith to Ft. Resolution on Gt. Slave Lake. They brought it down to Ft. Norman for us for $100.00, a reasonable sum. We arrived here on July 5th at 2 A.M., landed all our stuff at once, got our york boat cleaned out – it had brought a team of dogs all the way from Smith! – and loaded our stuff into her. Then came the job of arranging for Indians to track us up to Gt. Bear Lake.

Indians are here now, and we tried at first to make some arrangement with them, but they would not consider any proposition. They don't want us to go up there at all, and we finally had to give them up and try to get some local men. You have no idea of the difficulty of dealing with these men. The Crees were bad enough – the Chipewyans were worse than the Crees, and the Slaves worse than the Chipewyans, but these Dog Rib and Bear Indians are the worst of all. We require ten men. I can only get six, and only these by all kinds of arguments and inducements – roughly 50 "skins" – about $18.00 per man, and $21.00 for the steersman. I must find them food, utensils, footwear, tobacco, and feed their families while they are away. It is going to be an expensive business, but we are absolutely at the mercy of these people. The Hudson's Bay man has I believe done the very best he can for us, and he has carried the negotiations through. Unless these fellows find some fresh obstacle, we shall get away tomorrow, so I write tonight. It is undoubtedly a bad river to get up, and it will take 8 or 9 days. I almost regret our decision to winter at Dease; we could have been on the lake almost now by canoe. In fact, I had a canoe trip equipment already picked out today, for it looked as though we should either have to do that or spend all the first summer getting to the Dease.

I have met Hornby and Melville; the former is a very peculiar man.

I really don't know what to make of him. He is going back to the Dease River this summer and at first was very anxious to come with us. He did a good deal of "butting-in" in the various schemes we have tried to carry through; it has been a most involved business, and I have not time to go into all the factors that have increased our difficulties. Melville is the typical reserved Englishman; he seems a good sort, and I think Hornby and he can't agree. He is probably sick of Hornby. Hornby is collecting ethnological work for the British Museum, I believe, and he says he is going to try and winter with the Eskimos. Our york boat, the "Jupiter Ammon", is anchored about a mile up the Bear River, and we are ready for a start tomorrow. I hope we get off; losing time like this is too bad, but at least we are several days ahead of the time we expected to leave here.

We have not got any man to stay with us through the season yet; a man is coming this way from Dease River now; we may be able to get him. We could not find anyone suitable all the way down the river.

Well, we are in good shape, well equipped and all in good health. I shall write again from Bear Lake for the Indians to take back. Please excuse this hasty note.

Love to all.<br>
Always your affectionately,<br>
George M. Douglas

Sat A.M. Our Indians have found us; we are making a start today. Goodbye. G.M.D.

Ship Jupiter Ammon on Great Bear Lake, July 14th, 1911

Dear Cousin James,

I am glad to be able to report our progress so far. The journey up Bear R. I always considered one of the difficult parts of our trip and we have accomplished it safely. It was certainly a most arduous job: we were nine men all told and could not muster enough for turn about on the tracking line. Seven men was the most we could put on at bad places, one man to steer and one man in the bow with a pole. Our Indians worked splendidly once they got started: they stayed right with the job.

The river is very swift and mostly shallow. There were bad rapids where the river runs through a range of mountains and this was the part of the journey I dreaded. The ice piles up high along the banks here. Simpson at times had a bad job getting up it too, then rarely in

George Douglas in the *"Jupiter Ammon"*

north. Franklin also had hard work and could not ascend until nearly the beginning of August on account of ice. The water is very clear: a truly beautiful river and we made fairly good time to the rapid. For three miles the ice was piled up in a regular cliff some places over twenty feet high and overhanging worn out by the water. We tracked right along the foot of the ice: it overhanging us sometimes by eight or ten feet and we didn't stop to consider what would happen if a chunk should break off. One thing about these Indians, they seem to have confidence in the white man and will do whatever they see done by them. We had to lighten our boat half way up and sent two canoe loads of stuff to the head of the rapid, then we were able to get the Jupiter up the rest of the way. With ice at the bottom and rock shores at the top, that passage was certainly a thrilling one. The river is very swift all the rest of the way up and this

Comtemplation on the Bear River

tracking up is certainly the hardest work any of us ever did, but we reached the lake all right today and a welcome sight it was. I write this as we are sweeping the boat across Little Lake where Franklin had his winter quarters in 1826–27. The Indians are going back to Norman in a birch canoe they brought with them and will take this along. We propose to refit: we have to rig up a sail and we shall start in a day or two for Dease Bay. Bear Lake looks very lovely, wonderfully clear water. No ice in sight though the water is 41° in this bay. In the river it was nearly 36° and chilly work up to our middles every once in a while getting the Jupiter up. I had two sets of waders but didn't dare use them or the Indians would have quit if they had seen us do anything different from them.

Well, we are no worse the wear for the experience. Thank goodness we had plenty to eat—our bodies were the greatest machines for converting so many heat units into so much energy. The mosquitos were bad at the end of the river.

The banks were very pretty: many flowers and more thickly forested than I expected to find them. There seems to be a fair amount of wood at this end of the lake but not within a mile or so from the shores.

If all goes well we should reach Dease River by the beginning of August at latest: with luck by July 22nd. We may keep one of our Indian crew: he wants to stay with us to the head of the lake anyway.

We may be able to utilize some "house" that a trader put up: if not Lion and the Indian will build one. Sandberg and I propose to start out as soon as we can. We shall start early next season for the Coppermine and should get there early in May. We should get back to Ft Norman by canoe from Dease R. not later than Aug 25th. If the H.B. Co steamer has not left at that date we shall come out by her as far as Fort Smith and the rest of the way by our own canoe. Should we miss the steamer and yet arrive at Ft Norman before Sept 1st we should try and get out via Fort MacPherson then Peel Portage, the Porcupine R. to Fort Yukon. I should not attempt this route from Norman later than Sept 1st,

If we arrive later than Sept 1st we shall winter at Fort Simpson or perhaps Fort Smith, perhaps coming out by dogs when the freeze up comes. We are not prepared for more than one winter at the Dease.

July 14th "Little Lake' off Bear Lake

Here we are anchored in this pretty place where Sir John Franklin wintered in 1827. I have arranged with one of the Indians to come with us to Dease R. and spend two months anyway. His wife and child came too so we are quite a family. He will help Lion to build our winter quarters and so Dr. and I will start out at once [when] we get to the Dease. He cannot

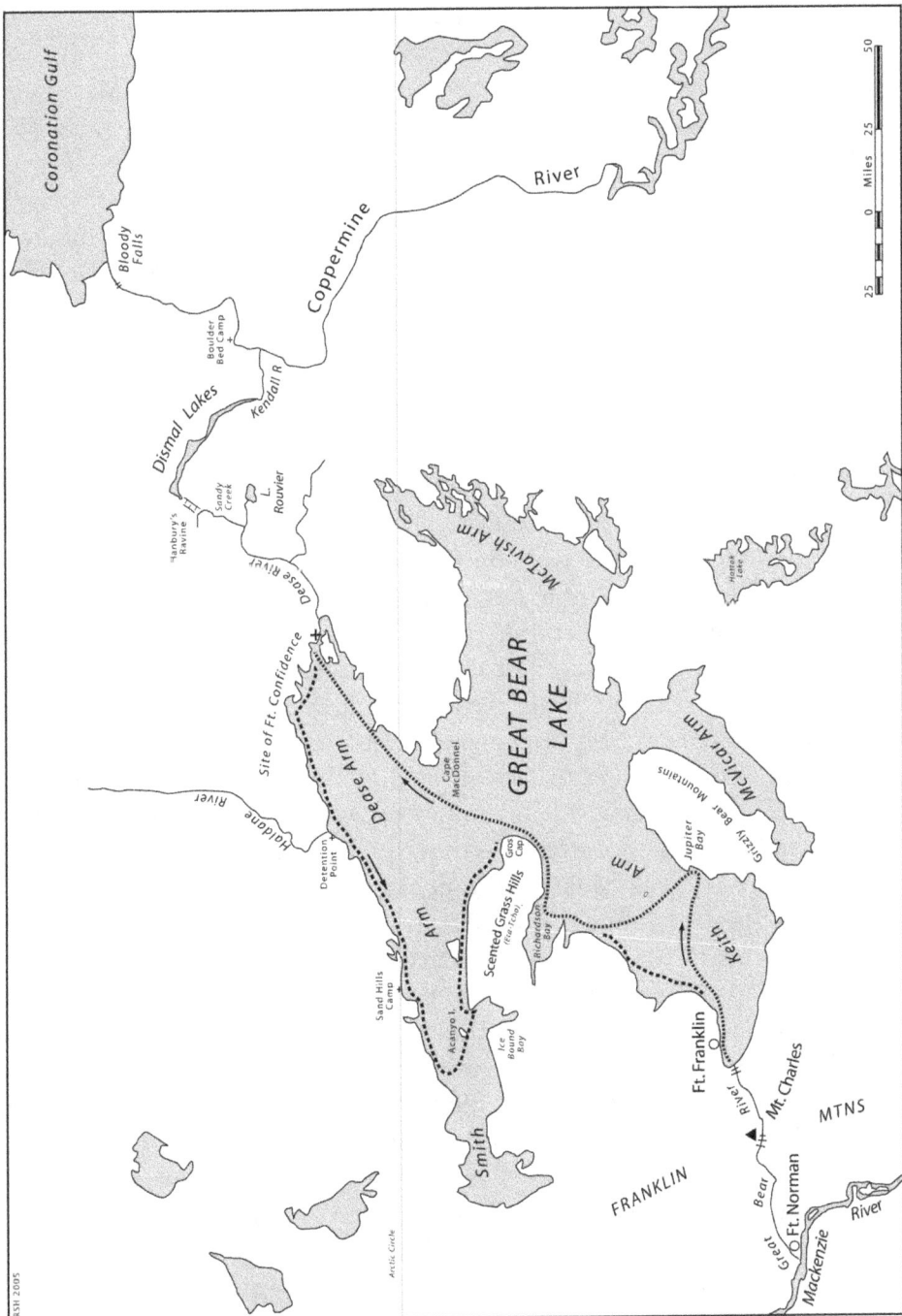

talk English nor French but his wife understands a little French.

It is a lovely evening and our Indians go back tomorrow early so I must close.

Love to all

Always yours affectionately

George M. Douglas

Hodgson's Point, Dease R. Gt. Bear Lake

December 26th, 1911

Dear Cousin James,

We may have the opportunity to send out letters by some Indians who are now hunting on the north shore of Great Bear Lake, and who may go back to Fort Norman shortly. So in the hope that it may reach you safely, I shall write a short account of our adventures. I have to condense things, and to leave a lot of "loose ends", but dispatch and delivery are uncertain and a fuller relation must be left for some better occasion. I expect you will have received Lion's letter written from the Dease R. after the Dr. and I had started for the Coppermine, and will have heard about our trip across Great Bear Lake. We arrived at the Dease R. in good time, though we were a very short–handed crew and dependent altogether on fair winds. But we went all the time: Lion took one watch and I took the other, the Dr. was cook and François, the Indian, ordinary seaman, and either slept or looked wistfully at the shore: such navigation was unprecedented in his experience. Fortunately for us in our crazy old vessel, we had fairly good weather and made the trip in eight days.

An old factor of the H.B.Co., Hodgson by name, had spent last winter near the mouth of the Dease R. hunting and trapping. We met him on his way back to Fort Norman and got his permission to use the shack he had put up. It proved to be a very poor thing but came in useful for a store house and we decided to build a cabin for ourselves at the same place, about 2 miles up the river. We called the place "Hodgson's Point".

We were pleased to find fairly good sized spruce trees here, not expecting anything but the scrubbiest of timber. There was much to do on our arrival but got things fixed up in a few days and on July 28th made a start up the river.

We went provisioned for 6 weeks; the total weight of our equipment was a little over 600 lbs., and everything of the very best. It had cost a lot but events proved its value: only the perfection of our equipment pulled us through. The Dease R. is a shallow rapid stream not much larger at Hodgson's Point that your East River in the Adirondacks. There are a few quiet stretches where the river flows through sandy plains. Tracking was impractable in such waters and with only two men: we simply had to wade the canoe up, it was hard work and slow progress.

For guidance we had Hanbury's very detailed account of his trip from the Dismal Lakes to Great Bear Lake: Simpson's account of his journey (very unsatisfactory): and a map made by J. M. Bell of the Canadian Geological Survey purporting to be a "paced survey of the country between Great Bear Lake and the Coppermine R". This latter was absurdly inaccurate and as it did not check up with either Hanbury's or Simpson's account of the country, we suspected as much before we saw the Dease R. ourselves.

I enclose a sketch map, a compass traverse I made myself, it does not pretend to give more than an approximate idea of the country but it will enable you to follow our course.

It was the morning of the 3rd day (31st July) after leaving Hodgson's Pt. when we got to the forks of the river, navigation had been difficult, a mere succession of shallow rapids wading nearly all of the time, lightening the canoe frequently, and occasional portages. According to Bell's map, the full absurdity of which we had not realized, we were now only some 15 miles in a straight line from Dismal Lake. So we camped at a prominent hill close to the river, we we called "Observation Hill", to investigate a little. We got a very extensive view to the north, a wide plain dotted with lakes and muskegs and sometimes gravel ridges and our view to the north was bounded by a range of mountains some 40 or 50 miles away. These proved afterwards to be the divide between the headwaters of the Dease R. and the Dismal Lakes but we did not know that at the time. I may say that this is a very puzzling country and especially difficult to trace a river course. Again and again we were baffled by the difficulty of distinguishing lake from a quiet stretch of river or mistaking a series of lakes for one large one, or the reverse.

Simpson's account was too vague to be of any use to us so we depended on Hanbury for a route over the divide. If you happen to have his book you will see that his account of the journey from Dismal Lakes is very detailed, and on the whole we found it accurate except in one important place. We knew that we would strike another fork in the river where his "Sandy Creek" came in and these forks we were prepared to identify by a "Kopje–shaped hill" described by him. We had also expected by various references and directions that Sandy Creek came in on the east side. [A map from Richardson and Rae's Franklin search expedition shows the relations correctly. It is surprising that Douglas did not have their map. Richardson, J., 1851, Arctic Searching Expedition: A Journal of a Boat–Voyage Through Rupert's Land and the Arctic Sea in Search of the Discovery Ships Under the Command of Sir John Franklin: Longman, Brown, Green, and Longmans, London, 413p. Richardson, J., 1851] We supposed from Simpson's account that the main branch of the river was the most westerly, he expressly stated so and his route had been by that.

Three more days of hard work in bad weather brought us to the junction; there was Hanbury's Kopje–shaped hill, impossible to mistake, about a mile above it. But the small branch was the most westerly and the main river from the east. In size the westerly branch answered to Hanbury's description, but in characteristics, the easterly. We camped at the junction and followed up each branch on foot for several miles and although we remained as much in doubt as ever, we decided to try the west branch. It was a small sandy rivulet; in some places scarcely enough water to float the canoe empty and no rocks anywhere. We went up for about five miles, poor navigation, but not so bad as some of the Dease R. We concluded that this could not have been Hanbury's route, it seemed to me also that we were going too far to the west. To add to our perplexity this small stream had branched about 2 miles from the main river. We went back to the little junction, camped, and decided to strike out due N on foot and see if we could find where the Dismal Lakes really lay. We followed up a high ridge of granite, "Granite Ridge", to a point about eight miles due N of the junction. The ridge ended here, some 8 or 10 miles further on lay a range of mountains and the country between very difficult, many lakes. But from the end of Granite Ridge we could trace the east branch to a large lake and decided to try that way. It took us most of two days to get within 6 miles of this lake and the river became so difficult that it meant a complete portage of everything for the rest of the way. A most forbidding country it was, the small stream fairly lost in a wide gravel bed, gravel hills ground up into irregular shapes — glacial moraine. The portage would have been a difficult one, and taken three days anyway, and then we had no idea how far we might still be from Dismal Lake or what kind of a route we might find. Judging by the considerable easting we had made we were still a long way to the south of them and the country towards the head of the lake appeared very rough. I was well nigh in despair, it was now August 7th. With the time at our disposal, and under our circumstances, this was evidently an impossible route, there remained nothing else for us to do but to turn back and to stake everything on the little western branch we had already attempted. I have marked the place where we turned back as "Camp Despair" and I don't want to pass a time like that again. Our chances for getting to the Coppermine that summer then looked desperate indeed and to pass the winter with the recollection of a baffled attempt was not a cheerful prospect.

We turned back and ascended that miserable little stream a second time, and only a few miles beyond where we had turned back on the first occasion we came on country that began to answer to Hanbury's description again; and presently the identification was complete. In describing it he has simply left off at a point some 6 or 7 miles above the junction a curious gap in an otherwise accurate description. We soon recognized that this must have been Simpson's route also.

The next three days we struggled on having to do a great deal of portaging. At "Simpson's Point" we had to make a complete portage of 2 1/2 miles and several others of 1/2 a mile and less. The stream got smaller and smaller, we finally camped at a place only a mile or so from its source in a miniature glacier, and near what we thought might be Hanbury's route over the divide. We started out again on foot and found that we were sure enough on the right track this time. The "ravine" described by Hanbury was unmistakable, and beyond this lay a narrow tongue of water among mountains which we knew must be the end of Dismal Lake. The actual distance across the divide was 6 1/2 miles; six or seven small lakes or ponds lay more or less along the route and they helped us by nearly a quarter of the distance.

We had camped in the afternoon and had started out that same evening. Just as we were about to turn back at Dismal Lake I caught sight of a figure through my glasses while spying around for deer. It disappeared behind a hill on which I could see some kind of a little camp, and thinking it might be an Eskimo, and possibly timid, we made a cautious approach and came on him before he had time to get away. He was an Eskimo all right, and at first he was greatly frightened, fairly scared to death. We held up our hands and shouted Teyma (about the only Eskimo word I know). He was soon reassured and seemed quite inclined to be friendly. Apparently, he was alone, we could see no signs of any others, he had a small roll of skin supported on some crossed sticks, a bow in a skin case, some arrows and spears. We made signs to him to come along, we wanted him to help us on the portage. He came along uncertainly, followed us at a gradually increasing distance, finally turned back and made for his camp, took it down, and disappeared behind the hill. He was an intelligent looking chap much more prepossessing than the Indians and we were sorry that we could not induce him to stay with us. Probably it was the first time he had ever seen a white man, he was certainly much frightened at first and though we did all we could to make friends with him, he was not altogether at ease.

Two days of strenuous work took us over the portage and on Aug 14th we started down Dismal Lake. I had hoped to be there by Aug 12th, so we were not so much behind after all, in spite of the time we had lost around Hanbury's Kopje.

The west end of Dismal Lake certainly looks its name, on the most westerly lake I should say, for there are in fact 3 lakes of slightly different elevations connected by narrows. We had had bad weather ever since

Observation Hill. This is a miserable climate, we had incessant light rains, raw NNE winds and frequent mists. And this bad weather stayed with us.

We went as far as the first narrows in one day, There were many traces of old Eskimo camps here and one apparently inhabited; a kayak carefully put away, some skins, a paddle, and some spears, and some fishing line of hide with a copper hook. We left a few trifles around the camp, and camped a couple of miles down Kendall R. the second day after leaving the head of the lake.

I hope to give you a fuller account of our trip some time. I find myself running on at great length yet having to omit much.

The Kendall R. was unexpectedly difficult. Hanbury talks of being able to run down it in a couple of hours with a good head of water. We evidently had a much higher stage of water than he did. The river is a swift boulder strewn stream, very difficult to run in a canoe as we found it. Running down such a river is always nervous work and I must say I began to feel the strain a little. The first part of the river is particularly unpleasant and we had pouring rain to add to our troubles. The descent seemed interminable, we had been running without interruption for nearly seven hours before we got to the cañon that I was prepared to find near its junction with the Coppermine R. Hanbury describes it, and Simpson gives a somewhat exaggerated account of it. It is a small "box" cañon through a limestone ridge with ugly enough walls. We landed above the entrance. I examined ahead and picked out a course that I thought we could run without undue risk. The first and swiftest part we came through all right, but we struck a boulder near the bottom that smashed a big hole in the canoe and threw us off our course into an ugly "set" against the wall. We slid past those rocks with not more than two inches clearance and came out all right at the bottom though in a sinking condition.

The Coppermine is a much larger river than I expected, larger than the Great Bear River and very swift. But good water as far as we went. We were delayed a day at the mouth of the Kendall R. while I fixed up the canoe again, but made a start on Aug 18th and with that swift current we ran down to the Coppermine Mountains in a few hours: the pleasantest part of all our trip. We made camp at a little stream described by Franklin and called by us on the map "Stoney Creek".

That first day at the mountains was a lucky one for us. Our food supply had been getting low, and on our arrival we had only 12 days grub left. Our consumption was far in excess of what we had figured. When engaged in

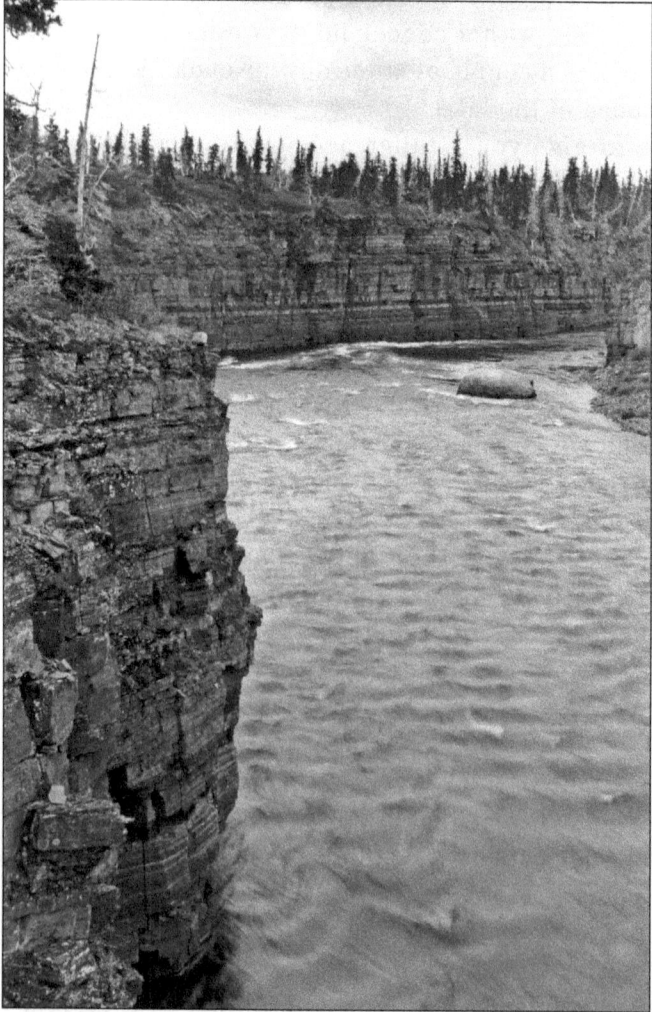

Cañon of the Kendall

such a heavy work in a cold raw climate and often wet, one eats a great deal: it is purely a question of thermo–dynamics, and from experience I would just as soon credit a man doing really hard work and heavy work on short rations as I would believe figures for any mechanical test giving absurd efficiences. Of course one can go on short rations, but one's effectiveness falls off fearfully and rapidly.

We had got duck, young geese, and ptarmigan on our way and a few fish, but we did not try much for the latter till we got to the Kendall R. We had only seen one caribou so far: I had shot a big bull while hunting around for a route on Granite Ridge. We could take only a little of the meat, but it was very welcome. Of musk–ox, which we expected to be our main stand by, we had seen no signs whatever, and as far as we can learn from Indian and Eskimo sources the animal is practically extinct here now. On the Kendall and Coppermine R. we made strenuous efforts to get fish with with nets and lines but got only one lake trout and one whitefish. We had the net out in several places on the Coppermine R. but caught nothing there. With us it was always a question of pushing on; we could afford no time to hunt and had to be strictly content with what happened to come our way. And with the heavy work before us of ascending the Kendall R. and crossing the divide, I did not feel inclined to take chances on the food proposition. But fortunately for us the first afternoon I shot another bull caribou—the only one we saw, for the deer had already gone further south; it was providential. I had often reproached myself with extravagance in the matter of arms but here was where we got the advantage of the best: our very stay at the Coppermine Mountains depended on the success of a necessarily long and uncertain shot. Well, so it was all along on this trip, we had tackled a job more difficult that I expected, especially for only two men: our equipment for the purpose was perfect and proved to be the last detail.

We spent until Aug 28th in those mountains. I asked the Dr. if we would care to enclose any kind of a summary of what we found, he preferred not to at present and I think he is right, it is premature and I don't know what may become of the letter, better wait until our return. I shall just say that we did as much exploration as that limited time permitted in a

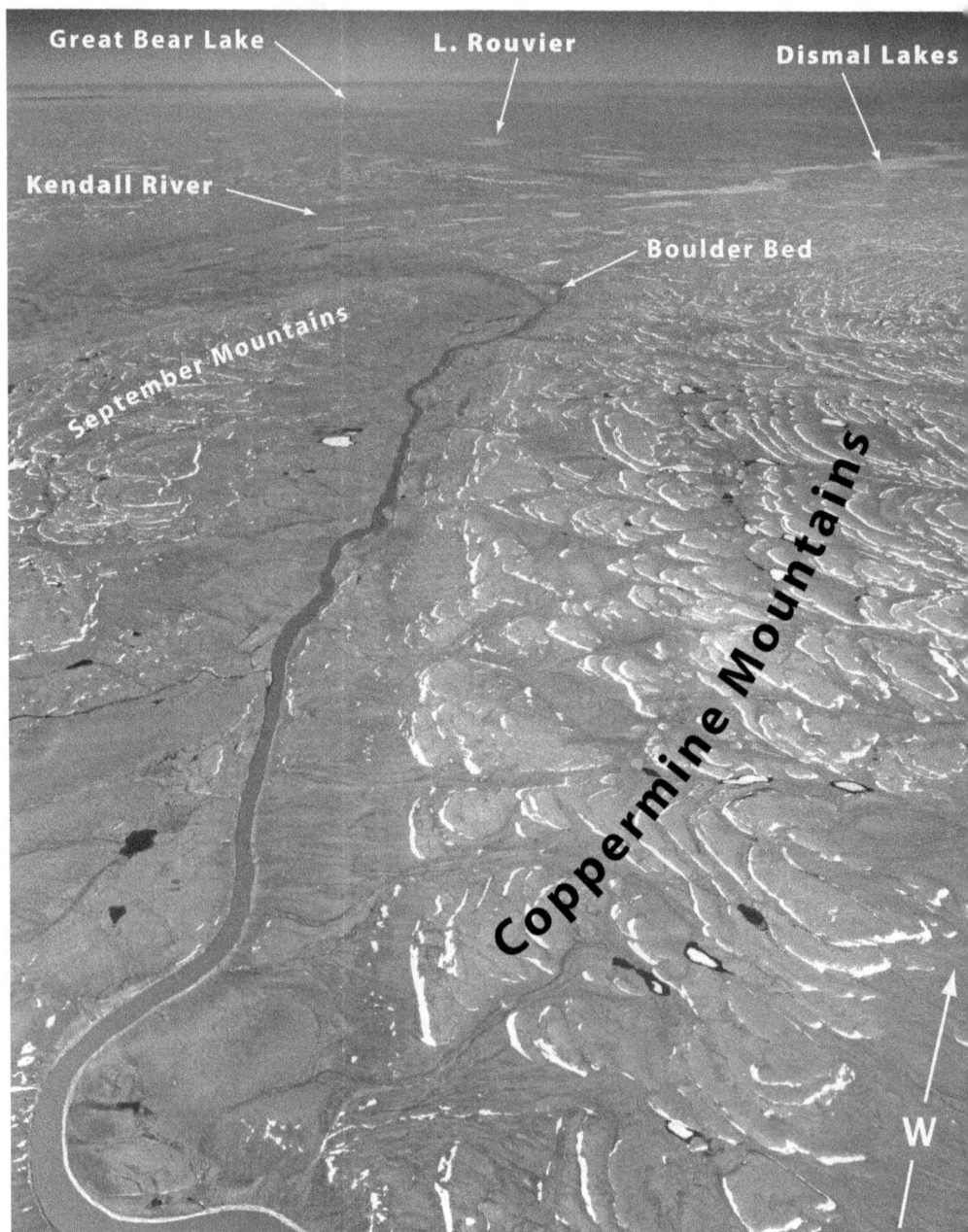

Looking eastward over the Coppermine River to Great Bear Lake. Snow drifts along south-facing cliffs accentuate individual lava flows of the Coppermine River Lavas

territory far larger than we had expected.

We found no native copper though we expected to do so almost any time and in many places. The mountains are a series of basaltic flows—the lower part a hard dense rock, the upper part amygdaloidal in structure, the "amygdals" [amygdules are filled vesicles or gas bubbles] containing minerals of quartz, calcite, epidote and chlorite. We found copper stain in many places and at "Glance Creek" we found a quantity of glance, there was much float in the bed of the stream and we traced this to where we found the rock in place. The stream had cut through near the top, it was much obscured by glacial drift and hard to tell its extent. We found beds of conglomerate at various places and at various heights but no copper in it. The same formation extends in every direction to the west, north, and east of our furthest point in those directions. The strike is N 60°E magnetic. We could see 30 miles about to the north, not quite so far to the NW, and about 20 miles to the east from the furthest we went in those directions. The mountains on the east side of the river were not so high as on the west—about 800 ft above the river compared to 1100–1200 on the west side. We found conglomerate beds at various heights there also and limestone underlying at one place.

The short season was closing in, we had heavy frosts at night and the uncertainty of the Kendall R. ahead of us. We took four days to get back to Dismal Lake, in the short interval we had been away summer had gone, or at least such summer as this country affords, autumn was almost over and winter coming on fast. I was much relieved to get the Kendall R. behind us, it was a difficult job and I appreciated Hanbury's relief as he coiled away his tracking lines on Dismal Lake and came to camp without the roar of rapids to remind him of what his next day's work would be. We had not the advantage of being able to track even, we simply had to wade the canoe up and the character of the stream, swift, with large boulders, made that especially difficult.

We stopped to make some investigations on Dismal Lake, the mountains on the south side are quartzite and diorite? but on the N side and as far along as the 2nd narrows we found the same formation as the Coppermine Mtns., the same mountain ridge in fact, they come right down to the lake. We found conglomerate beds here also but no copper.

We spent several days at "Glacier Cove" but were baffled by constant mists and rains from exploring more than a few miles. Our passage across the lake was a stormy one—winter was beginning as we crossed the divide with frosts and snow storms and the small lakes beginning to freeze up, we were none too soon. We made good headway and got back to the main Dease R. at Hanbury's Kopje on Sept 7th.

We had left a cache of flour and beans and a few things at the junction and found that Eskimos had been around there in our absence. They had left some stuff in it, a sealskin shirt, some slippers, and arrows. I thought it likely that they had found and appropriated the remains of the caribou we had shot near the place and this was their acknowledgment. Apparently they had left the country again and though we saw some smoke 8 or 9 miles up the river we did not stop to see what it was. We were anxious to see how Lion was getting on now and we got back to Hodgson's Point on Sept 11th.

We found Lion well: he had put up a splendid little cabin 14' x 16' with well-mudded walls and a good fireplace. It has proved a snug little house indeed.

It was much milder on the lower Dease R. than it had been on the Dismal Lake side of the divide, so to get all the benefit we could while the river remained open, Lion and I started up again on Sept 15th. We went partly to hunt and partly to explore. We went up the S. Branch to its source and made a pretty good guess where Bell had been and what he had taken for Dismal Lake, But we got no caribou, this was disappointing as we were depending on them for fresh meat, and the Dr. and I had seen a fair number on our return and had happened to kill them just as we needed them.

We went as far as "Stefannson R." and then turned back, snow had been on the ground for some days and the river was likely to freeze up any night, so we went back to our winter quarters, and the river did in fact freeze up two days after our return.

About Hornby and Father Rouvier I shall have to tell you some other time. When we saw them last at the Bear River it was their intention to winter at Dismal Lake, the Father to convert the Eskimos, Hornby to trade with them. I should say that both of them have spent a number of years in north Canada and they are experienced travelers. They arrived at the Dease R. about the second week of August and had started up the river. The Dr and I saw nothing of them on Dismal Lake and we began to wonder what had become of them, when Hornby appeared on the scene with some dogs packing his outfit. Like ourselves they had taken the east branch for a route to Dismal Lake and had got as far as the lake, "Lake Rouvier" as we called it and found the country too difficult and the season too late to go any further so decided to build their winter quarters there. Eskimos appeared on the scene in great numbers according to them, the father could speak no Eskimo and their intercourse was very limited. They put up a little hut on the shore of the lake and Hornby had left the father there and was going back with some more supplies. He said that the Eskimo were still around the lake.

By the 20th of October we had regular winter though not so much snow had fallen; the temperature was ranging from zero to twenty and twenty-five below. Lion and I decided to make a trip up to the lake with Hornby and see what we could do in the way of winter traveling. In that respect we are badly equipped as might have been expected from our lack of experience. Moreover we found it very difficult to get things like toboggans, fur clothing, etc in the North: it is surprising at first, but when you see how the people live up here and their general way of doing things, you begin to understand it. We were able to get only some very poor dog sleds and short heavy toboggans. It is a sure thing that one can do nothing in this country in winter or spring without dogs and these we had beforehand decided to do without and for various good reasons.

Anyway, the trip showed us what we were up against on a man hauling proposition. We didn't notice it much on the way up, having unusually good conditions for traveling and the help of Hornby's dogs. We got to the lake in 4 days but to our disappointment the Eskimos had all left. The Father was in good health and spirits but had decided not to spend the winter at the lake. He went back with us on our return as far as the kopje. Lion and I went off to the west on a hunt while he went straight back to Hodgson's Pt. He and Hornby are now living at a shack put up by Hornby and Melville last winter about halfway between old Fort Confidence and the mouth of the Dease R, but there were no deer in the country. Since leaving Lake Rouvier we had had very good conditions and had been getting on fairly well. After we left the kopje on our way home we struck snowy weather and then our troubles began, yet we had very light loads, only 110 lbs on one and 130 lbs, but our toboggans were no good for that kind of work—too short and heavy and very poor wood. Light as we were we had a hard job getting back to Hodgson's Pt. Considered as a draught animal man ranks poor indeed, and hauling a toboggan is a very exhausting job.

Now we are at the height of winter and living a very regular and well ordered life. We take turns as "cook", "wood chopper", and "hunter", week and week about and these occupations keep each one busy. We do not have more than a few hours daylight (or twilight), we saw the sun last on Dec 9th. The weather up to Dec 20th was very dull, not particularly cold, but always a little snow in the air, yet the total snowfall has not amounted to more than 2 feet so far, and very light snow at that. The last week we have had clear and cold weather, the thermometer below −40° most of the time: − 49 1/2° was the lowest so far.

Of game we have seen very little; between Sept 10th and Nov. 20th we got no caribou at all and ptarmigan was our main stand by. Then a bunch of caribou drifted into our vicinity and we got 6 in two days, but all small ones. However we considered ourselves lucky, for the Indians, of whom there were several families, and Hornby and the Father got none at all.

The Indians are now out on an extended hunt after caribou and I believe it is their intention to go back to Ft. Norman as soon as they get some meat.

Thank goodness we have plenty of grub, we would be in a hard fix if we had depended on living off this country.

The movements of the caribou are notoriously uncertain, they were abundant here last year but were hunted a good deal by Hodgson and his outfit and by Hornby and Melville, and they may give the Dease R. a wide berth this year.

We are all well and in good spirits, could you see us as we are we should probably look rather wild but we are used to each other's appearance now.

Our plans for next spring are indefinite, we have realized several objections to our first scheme of making the trip before the river broke up, hauling the canoe and stuff on sleigh, but we shall be guided by circumstances as you may be sure we shall do the best we can. We may try to make some investigations to the N of Great Bear Lake, the Indians report copper there and on our way across the lake we found some coal on a beach where we landed which François said came from a hill on the north shore near the Haldane R.

We have heard nothing of the outside world since we left Edmonton. There may be mail for us at Fort Norman, at present I see almost no chance of getting it until our return there next summer. This is the only letter I am sending out.

Goodbye for the present and with love to all.
Always yours affectionately,
George M. Douglas.

Hodgson's Point, February 25th, 1912

By January 19th there was no sign of the Indians and we had concluded that they must have hunted so far to the W that they had gone on to Fort Norman. The winter mail packet leaves Fort Norman early in February, the Father had some mail that he was anxious to send out by it so Hornby decided to make the trip to Ft. Norman himself. The Dr. was also anxious for a change and to have some experience of winter traveling so he and Hornby made a start on Jan 20th. Hornby had only two dogs and we were scarcely surprised to see them again in a week. They had got so far as Cape MacDonnel and decided that the traverse across to Gros Cap was too much for them with their motive power, so came back again.

A couple of days after they had started, the brother of François, the Indian who came with us on the "Jupiter", came in. The Indians had had a bad time, some of them lost their dogs by starvation and, with the exception of François and his family, had made the best of their way to the little Indian village on the other side of the lake. François was camped a couple of days journey to the W on the N shore of the lake and had sent his brother in to get some grub and ammunition. We gave

Sleeping quarters at Hodgson's Point

them some and François came in again yesterday, he had struck better luck and had killed caribou. They are going to start for Ft. Norman again in a day or so, Hornby is going with them and leaving tomorrow morning so I shall send this letter out by him.

Our plans for the spring have taken more definite shape now. François and Hornby expect to return from Fort Norman in about a month. François has four dogs, Hornby now has three and we shall get as many as we can in Fort Norman, four if possible, but this is unlikely. I have made a definite agreement with Hornby to work for us in the spring for 2 months or so. I shall get François to haul our stuff up to Lake Rouvier in April and make a start ourselves about April 20th. With Hornby's aid we hope to lay 3 caches: on Dismal Lake, on the Kendall R., and at the Coppermine Mtns. We shall take what we require as far down the river as we want to go (I hope to the sea), the Dr. and I shall work our way slowly up the river while Lion and Hornby are hunting, making caches, etc. I think we may expect to be able to see a good deal of the country by the end of May, but at least this will give us a chance to see all we possibly can and to take advantage of all the time possible between spring and latter part of July. I hope to get the last steamer from Fort Norman towards the end of August. This will take us to Fort Smith, the rest of the journey we shall have to do in our own canoe.

I hope neither Hornby or François go back on us, if they do it will be a case of going by canoe again, which will not give us much time though there are advantages in that way of traveling and the season being earlier on this side of the divide than it is on the Coppermine still gives us the chance to be on the ground at the best time, or rather at the <u>best</u> <u>part</u> of the time.

We have been going on a very regular way of living here. January was a cold month, –58° was the lowest we had, the observed mean for the month was –31 1/2°, the true mean was probably nearer –40°. We are all well and consider ourselves fortunate that we came well provisioned for though we have been hunting all the time, we have seen no caribou since November and some efforts at fishing through the ice on the lake were not successful. On the whole this experience of wintering has been interesting and not unpleasant, the Dr. perhaps doesn't take quite so kindly to a life of routine as Lion and I do, one must take an interest in trifles: a new kind of stew, a better way of scouring a pot, or the find of a good stick of dry wood, must be events.

I hope Hornby will bring us back some mail from Norman.

With love to all,

Always yours affecty

George M. Douglas

# George M. Douglas

Fort Norman, August 17th, 1912

Dear Cousin James

I expect you will have received the letter I wrote from the Dease R. in February telling you of our doings last summer and autumn.

This letter was sent out by Hornby, who made the trip to Fort Norman with some Indians leaving the Dease R. on Feb 28th.

Only two or three families of Indians had been in our neighborhood for the winter and they had spent most of the time further west along the shore of Great Bear Lake. It had been a hard year for them, they had starved in the early part of the winter and got no fur at any time. They had only about six dogs between them all but with Hornby's two dogs they mustered enough to haul two toboggans with a prospect for a successful trip.

March was a cold month: hitherto winds from the west had always brought a quick rise in the temperature, now we had strong westerly winds and temperatures of −40°, even −50°.

A few small bands of caribou passed in our vicinity so Lion and I made a trip up the river about the middle of the month. We had no dog food for our projected trip in the spring. So, in spite of resolutions to the contrary made after our last autumn's experience, we hauled a toboggan up the river again. We got no caribou and saw no further traces of them after a few miles from our shack. The trip was only a short one anyhow, but interesting rather on account of the low temperatures we had.

On March 24th the Indians returned from Fort Norman, but Hornby had remained behind. They brought instructions from the Mission for Father Rouvier to return and had sent a team of dogs to bring him back. One of the Indians was going back with him. Hornby intended to wait till they got back and return with this Indian.

They brought us our mail, the first news of the outside world we had received since leaving Edmonton in May and this was up to the beginning of November. On March 27th the Father left for Fort Norman, he had spent a good deal of the time with us after Hornby had left and we were sorry to lose him. I had made arrangements with François to haul some stuff up to Lake Rouvier as soon as the Indians got back. His brother Modeste, who was one of the party that had just returned from the Ft. Norman trip, also agreed to make one journey up to the lake. We intended to find some route over the divide to Dismal Lake from there.

So we had two teams of 3 dogs each and on April 4th the Dr., François, and Modeste made the first trip with about 400 lbs of stuff. They had good weather and good traveling conditions and made the round trip in 7 days, which was pretty good work.

Then François backed out of agreement and refused to work any more, we never learnt why, probably no reason at all; these Indians are so thoroughly unreliable that for them to fall down on you is the only thing that can be relied upon to any certainty in your dealings with them. But after considering the matter for a few days, François agreed to lend us his dogs so the Dr. and Lion started up with another load. They had good weather, April is the best month of the year to travel in this country, the snow is harder, the temperatures are moderate, the days long and generally fine. Towards the end of the month it began to get decidedly milder. There would be signs of thawing in the sun sheltered spots. April on the Dease R. may be compared very well to February in eastern Ontario and Quebec. On April 15th the maximum temperature was +33°, the first time the thermometer had been above freezing for 177 days. But it turned cold again till the beginning of May.

On April 17th Hornby got back from Fort Norman. He didn't want to work for us now under any kind of agreement but was willing to do what he could to help us out. I think he had got some kind of a commission from the H.B. Co. to investigate the possibilities of opening up a fur trade with the Eskimo of Coronation Gulf: or wanted to do some fur trading with them also on his own account. He didn't want to make the trip by himself, in fact he wasn't very well fixed up for it so it was for our mutual benefit to make one party.

He had brought us some of our mail from Fort Norman, also news from the H.B. Co that the last steamer would be leaving Fort Norman on Aug 1st.

Lion and the Dr. got back from their trip shortly after Hornby's arrival. Hornby wanted to take up a load of trade goods for the Eskimos and we had still a good deal of stuff to go up so he had to make another trip up to the lake and back again. The Dr. remained there this time. Hornby, Lion and I made the final start from Hodgson's Point on April 30th.

We had mustered 7 dogs by this time. Hornby had been able to get only 2 more at Fort Norman, we borrowed two from the Indians and another we "rescued" from them in the last stages of starvation: poor old "Jack", he afterwards became one of our best dogs but in spite of good feeding he had just about strength enough to follow us the first day.

Hornby also took an Indian boy with us who was just about as much nuisance as help and who put away more grub than any of the party besides.

We had provisions for 4 men for 50 days, but no dog food, we had to take chances on getting caribou, the Dr. and Hornby had seen some on their last trip but had not got any.

We had good traveling and reached the lake on May 2nd, we came across some caribou on our way and had got 4, a good thing for the dogs and Jack picked up wonderfully.

So we had all our stuff up at Lake Rouvier anyway our plans were then to take a load of stuff across to the head of the Kendall R. (it required two trips or 4 toboggan loads to move all the stuff). Lion and I were going to hunt and prospect for a good route while Hornby brought the second loads. We expected to have our second base camp established in the Coppermine Mtns. by May 12th.

We intended then to make a start down the Coppermine R. soon as we could and move as much stuff as we could take without relaying to the second range of mountains, which we knew crossed the river some 30 or 40 miles below the Coppermine Mtns. We thought that these might be basalt also and the best field for investigation would lie between these mountains and across the Coppermine Mtns giving a cross section of the country. From this lower mountain range we intended to make a quick trip to the sea and back. Hornby was then to follow his own devices and either return to the Dease R. or stay with us while we worked our way slowly across the country again.

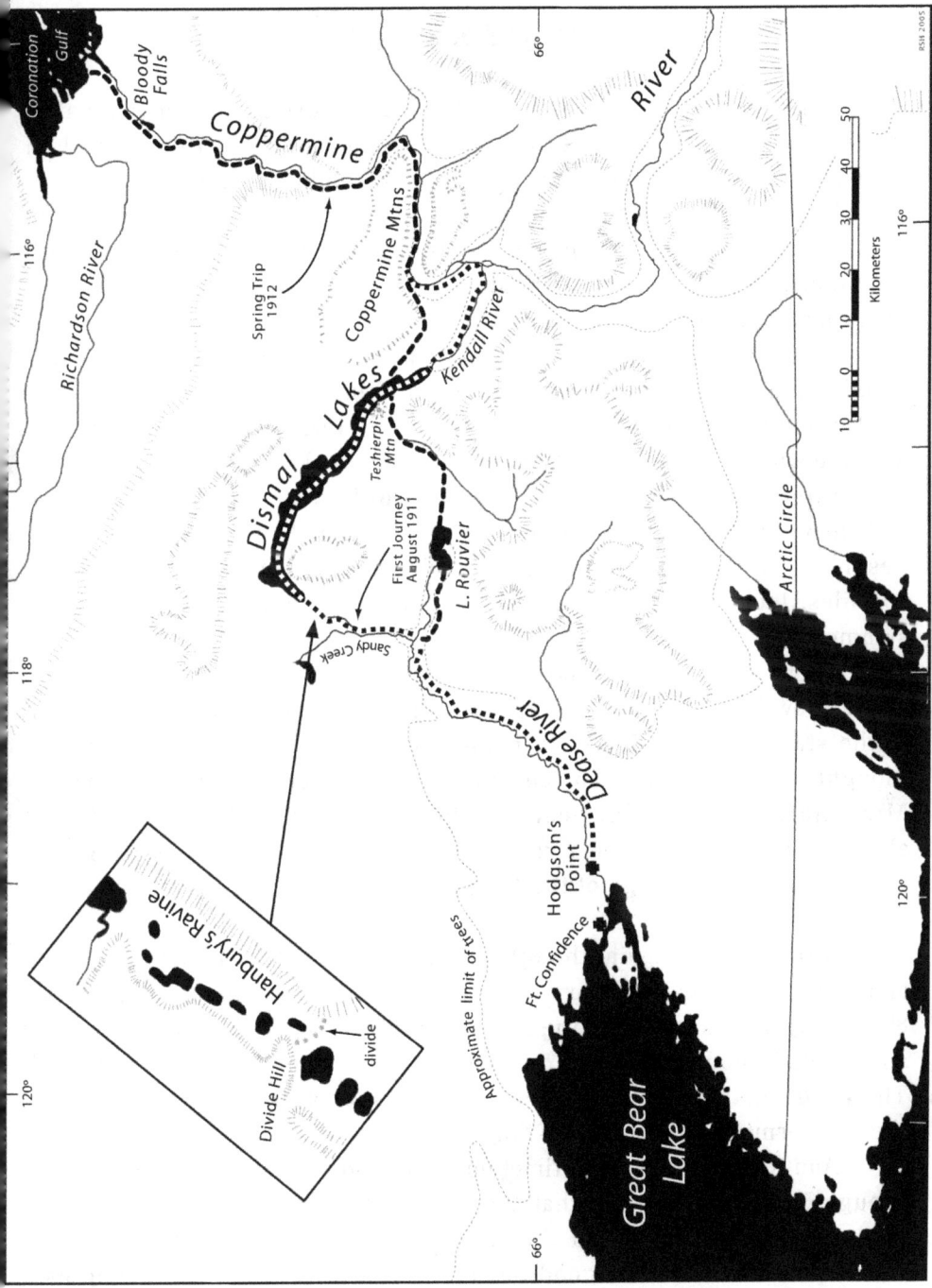

Coronation Gulf

Bloody Falls

Richardson River

Coppermine

River

Spring Trip 1912

Coppermine Mtns

Kendall River

Dismal Lakes

Teshierpi Mtn

First Journey August 1911

Sandy Creek

L. Rouvier

Dease River

Hodgson's Point

Ft. Confidence

Approximate limit of trees

Great Bear Lake

Arctic Circle

Hanbury's Ravine

Divide Hill

divide

Kilometers

10  0  10  20  30  40  50

66°

116°

120°

118°

120°

116°

66°

RSN 2005

We expected to walk back to Lake Rouvier fording the Dismal Lakes at the 1st narrows to descend the Dease R. in a canoe that Hornby had on the lake, returning to our house about the beginning of July.

How these plans worked out you will see. I make a note of them as it is interesting to compare what has actually been accomplished with what was intended to be done, though I have seldom known any satisfaction resulting from such comparisons.

On May 4th we made the first trip across the divide between L. Rouvier and the Dismal Lakes. We knew nothing about the route and after the Dr. and I turning back from it last summer, I was curious to see what it was like and where we would strike the Dismal Lakes.

It turned mild suddenly on the night of May 3rd and the next day we had heavy hauling but got across by a fairly good route and struck the Teshierpi R. about 5 miles from the Dismal Lake. We had come nearly 20 miles: lucky indeed that the Dr. and I turned back when we did last summer.

There is very little snow on those hills in the barrens, the high winds blow it off and this unexpected thaw decided us to bring the rest of the stuff to the Teshierpi R. before going any further for at worst we thought we would have an ice surface from there to the Coppermine Mtns. Hornby and the Dr. went back for the second load. Lion and I went ahead a little to look out a route and decided to cut across to Stoney Creek following the edge of the Coppermine Mtns from where they strike Dismal Lake.

It turned cold again, the others had a good trip to the lake and back again. A couple of days of stormy weather delayed us but on the night of May 8th Hornby, Lion and I made the first trip across: the route proved a fine one and we had good traveling. Hornby returned at once for the second load. The good weather held and we were established at that familiar corner of the Coppermine Mtns again by May 11th.

And we got over just in time, not a day too soon for the next few days brought such a mild spell that the snow disappeared completely except in the big drifts, the ice in the river was flooded and further progress was impossible. This was good for us as regards prospecting, giving us a chance to see the country much sooner than we had expected, but it looked much as though our trip to the sea would not come off, for we had expected to be able to travel on the Coppermine R. until the end of May anyway.

By May 17th we had given up all hope of a cold spell that would freeze up the river so we could travel on it. Much water was running along the shores and although by Simpson's account the river did not break up for another month, yet it looked now as though it might be only a matter of a few days.

We started to build a canoe, breaking up the toboggans to make the frame and we had a light sail–silk tarpaulin that we intended to cover it with. A week brought no change in things, the weather was too cold for any further breaking up of the river and not cold enough to freeze things so we could travel.

By May 20th I had given up hope of the river breaking up soon enough to let us profit by the canoe even if it was a success, our tarpaulin was so light and the whole contraption so frail that we had some reason to be doubtful about it. The Dr. meanwhile had been looking over the mountains in various directions from our camp at Stoney Creek and from another camp that he and Hornby had made about 10 miles further down the river, packing down a little grub and a tent.

Hornby was now getting ready for his trip back to the Dease R. He had also given up all hope of getting to the sea and seeing the Eskimo.

We got him to stay with us till we could pack some stuff another 10 miles below the Dr.'s first "outpost" camp. Hornby, Lion and I then returned to Stoney Creek, Hornby was going back to L. Rouvier. Lion and I were going to pack another load down the river but at the last moment we persuaded Hornby to stay with us.

The help he gave us certainly was most valuable. You would be surprised at the load a dog can pack on his back, we had four dogs that could carry 50 lbs each and one that could take over <u>60</u> lbs.

We had got a number of caribou, the dogs had been well fed and we had lived on fresh meat ourselves so there was lots of food. We took all that everyone and every dog could carry and found the Dr. again.

We moved slowly down the river investigating as we went. We had no more snow or only in such a light fall that the ground was bare again in a few hours of sunshine.

June came in with fine weather and June 3rd found us camped just above Hearne's "Bloody Fall", this was as far as we intended to move our camp.

On June 4th we met with Eskimos right at the Bloody Falls (They are not falls but only a rapid). Several families were camped on the opposite side of the river, some of the same families that had been to Lake Rouvier last summer. There was an ice jam at the head of the rapids and two of the Eskimos crossed on the jammed ice floes. They brought a musk ox skin with them and we tried to find out what we could about them. They crossed back to get another musk ox skin, the ice jam carried away, and we could hold no further communication with them. Hornby and the Dr. returned to the camp. Lion and I walked right on to the sea, about 9 miles from the Bloody Falls, over level sandy grass covered plains with small lakes and muskegs. We struck the sea between the mouth of the Coppermine and Richardson Rivers.

The ice was very smooth and level, no snow left on it. There was no open space along the shore as with all the lakes now but the ice was frozen solid to the beach and nothing to indicate any tide. Possibly the ice is sufficiently elastic to take care of such slight rise and fall as there may be.

On our return we came across another Eskimo camp, the most pleasant looking lot we had seen yet.

These Eskimos are certainly a very interesting people, they received us courteously, showed no impertinent curiosity like the Indians (and most white men), they were affable and laughing and cheerful with quick ways and manners and a power of understanding. They did not recognize the names of Stefansson nor Anderson. They knew Lake Rouvier and called it "Ar-ping". I had an Eskimo dictionary in my notebook that I had made up of words Hanbury gives and from a French Eskimo dictionary by a Father Petitot, who had been a missionary to the Eskimos at the mouth of the Mackenzie R. [Father Emile Petitot was a French missionary, who traveled extensively throughout the region. He made no fewer than eight journeys to Great Bear Lake, several by way of the chain of lakes and rivers running from Great Slave Lake. His travels were mostly solo and involved traveling incredible distances, some by canoe, but mostly on snowshoes during the winter. His reports are superb and contain many observations on flora, fauna and indigenous people. See: Petitot, E., 1875, Géographie de l'Athabaskaw–Mackenzie et des grands lacs du bassin arctique: Bulletin de la Société de Géographie, 6th series, v. 10, p.5–42; p 126–183, p. 242–290; Petitot, E., 1893, Exploration de la region du Grand Lac des Ours, Paris, 469p.] They didn't recognize any of Hanbury's words and but a few of Petitot's, very likely my pronunciation was to blame. To consult my notebook interested the Eskimos greatly and when I happened to get out a word that they recognized it caused them the greatest surprise and amusement.

We tried to learn from them where they found copper and gathered that it was a long way off in the mountains to the SW, indicating the heavy pack necessary for the journey and their exhausted condition when they got there: we gathered from their description of how they find it lying around loose that it must be float. Though they still use many copper weapons and utensils, they also have iron to about the same extent and I imagine that they have quit looking for copper for some years now. The Indians are just getting over their fear of the Eskimos, the last two summers have made a great difference. The Indians meet the Eskimos on the edge of the barrens and trade with them: hardware of various kinds, such as these Indians can spare, knives, axes, files and pots, in return for sealskin boots and dogs.

Then Stefansson, Hornby and Melville have brought a lot of stuff into the country and these Eskimos show an extraordinary ingenuity in turning things to account. None of them showed the slightest fear of us and I am now rather puzzled to account for the evidently scared condition of the man that the Dr. and I met last year. However I must tell you what we saw of the Eskimos another time and get back to our journey.

We turned south again on June 15th. Our discoveries will be described in due time by the Doctor, but roughly there are three ranges of basalt hills: the Coppermine Mountains, a range about 30 miles below the Coppermine Mountains, and at the Bloody Falls. Below the Coppermine Mountains between the other ranges is sandstone with occasional dykes of basalt.

Other ranges of what is evidently the same formation extend further north, Cape Hearne and the islands in the gulf [Coronation gulf] are apparently basalt. [they are gabbro, the intrusive equivalent of basalt]

The ranges below the Coppermine Mountains are much lower and of a slightly different character of basalt, we found no trace of copper in them.

We found copper in several places along below the Coppermine Mts., mostly below Glance Creek, a description of this must be left for the present.

We left the Coppermine River on June 15ᵗʰ and followed the strike of the mountains to Glacier Cove and prospected over the mountains to the NW from there. But we found the lake so high that to ford at the narrows was impossible. The ice was still fairly solid in the lake but open along the shores, its condition was precarious and not at all permanent, to quote the bos'un in one of Marryat's yarns [presumably renowned captain, scientist, and author Captain Frederick Marryat, 1792–1852]; and as a strong wind might have left us in an awkward position we decided to lose no more time than we could help.

So next day the Dr. went as far to the NW in twelve hours while Lion and I built a raft. There is a little small spruce among the gravel hills at Glacier Cove, the only place where spruce occurs on Dismal Lakes and we were lucky to get some sticks for a raft just big enough to float one at time. We got a line across to the ice and ferried our stuff on to it. As soon as the Dr. returned that night we started out. We piled all the stuff on the raft, which we had shaped roughly like a sleigh, the united efforts of all the party and the dogs were just able to haul it across the lake. The ice was almost touching at a point there and after having served as a boat and a sleigh, we turned our "Dolphin" into a bridge and landed all safe close to Teshierpi Mountain. We crossed the mountain that same night and camped at our old place on the Teshierpi R. We rested up for a day here and came across some more Eskimos. We finally got to Lake Rouvier again on June 18ᵗʰ. Our cache of grub was all right, also the canoe that Hornby had left there last summer.

Lion and I took the canoe and a good load of stuff. Hornby and the Dr. walked with the dogs, following the high land to the S of the river, which is shorter and better walking.

We had cold and windy weather, the canoe was a big heavy one and in spite of the higher stage of water, the first twenty miles to Hanbury's Kopje was scarcely less hard work than when the Dr. and I came down that same stretch last summer. Our canoe was moreover in very bad shape, we had to repair her a good many times before we got through.

From Hanbury's Kopje the journey was pleasant, we reached Hodgson's Point on the afternoon of June 22nd, the others came in a few hours after we did. We had been 54 days on the trip, everything had gone well, we were all in the best of conditions, the dogs especially the best test of a trip. We always had plenty of grub and altogether the expedition had been very successful though I regret that we could not do anything on the east side of the Coppermine R., and I should like to have seen more of the country NW from the Dismal Lakes, but in a case like this, the more one does, the more one sees must be left undone.

The trip had been a pleasant one, far less arduous than last summer's trip the Dr. and I made. The larger party makes it easier and knowing the country is a great advantage.

We tried to get Hornby to come out with us but he didn't want to leave the country yet. He hoped to get some fur from the Eskimos and it was his intention to return to Lake Rouvier with two Indians. We gave him a lot of stuff: what provisions we had left, the "Polaris" rifle, ammunition, etc.

We had already arranged with Father Rouvier that he could have the use of our house at Hodgson's Pt. this coming winter.

We bade farewell to the Dease River on June 26th. It was our intention to investigate as much as we could along the N shore of Great Bear Lake, we heard rumours of copper having been found there, also what from the Indian descriptions might have been coal. We were provisioned for 30 days. Unless the country looked promising it was our intention to push on to Fort Norman and catch the first steamer if possible.

To describe our voyage across Gt Bear Lake properly would require a letter to itself.

We struck ice as soon as we left Ft. Confidence, but a strong NE wind has opened up a passage between the ice and the shore and though we had to get through or over ice jams in places, we made good headway and got to a point some miles west of the Haldane R. by June 28th. I have omitted to mention that we had the "Aldebaran", the big canoe we got at Edmonton, and we had a load of about 900lbs on board. The first twenty or thirty miles is mostly limestone formation. We came across slate in one place containing carbonaceous matter, there might be coal in that country.

West of the limestone formation, extending some miles west of the Haldane R. we found a great deal of basalt along the shore of the same characteristic as the Coppermine Mtns. It was evidently glacial drift and might mean that those mountains extend very much further west than we suppose. [they do] At Point Detention some five miles W of the Haldane R we were hopelessly hung up by ice until July 3rd. The mountains on the N shore were too far for us to get to, they appear to be two isolated ranges and apparently are not basalt.

Another N wind opened up a passage along the shore so that we got to a point opposite Grange I [?], "Point Traverse" where we intended to cross Smith's Bay, or Good Hope Bay, as the Indians call it. The ice was apparently solid across the lake, we waited there until July 7th and then attempted to go around the end of the bay but on reaching the NW corner the prospect looked so bad that we went back to a point opposite Grange I. again. On July 11th we had a NW gale that broke up the ice and we got across all right on July 12th. We were hung up again at "Icebound Bay" and another point some 20 miles W of Gros Cap. At this place we found coal: we traced a seam for about 1000 yards along a steep sand and along shore line. In some places where it was well-exposed, it measured 9 1/2 feet in thickness.

After we rounded Gros Cap our journey was uneventful and pleasant enough. We reached the Indian settlement near the site of Fort Franklin but no one was there. Hoping that the steamer might have been delayed so that there was still a chance of catching here we decided to push on at once for Fort Norman and ran about 40 miles down the Bear R. that night when we met the Indians and Father Rouvier and learned that the steamer has sailed a week before.

We arrived at Fort Norman on July 21st. Since the beginning of August we have been expecting the steamer anytime. Now we begin to think that she must have met with an accident somewhere.

Unless she shows up by Aug 21st we shall make a start down the river and attempt to get out by the Yukon.

S.S. Mackenzie River, Aug 22nd

The steamer arrived at Fort Norman on Aug 17th. She had been ashore in Great Slave Lake. So we shall arrive about the same time as this.

Your letter received at Fort Norman, also the magazines for which many thanks.

Athabasca Landing
Oct 16th, 1912

We arrived at Fort Smith on Aug 29th after a pleasant enough voyage from Ft. Norman. We went on direct to Smith's Landing and made the voyage from that place to Fort Chipewyan in our own canoe, the "Aldebaran". The H.B. Co. were sending out a scow from that place and we came the rest of the way with them, we might have got here a few days sooner if we had come straight on in our boat, but we were too heavy loaded to make that long journey up the Athabasca above McMurray. We left Chipewyan on Sept 13th, the H.B. Co. steamer took the scow as far as Ft. McKay below McMurray, then it was a case of tracking the rest of the way and though a rather interesting way of traveling, when you haven't got to track yourself, it gets rather tedious. It took 17 days to get from McMurray to the Pelican, a distance of only 135 miles but very bad water, especially at this time of year and I was thankful we did not attempt it ourselves.

We leave for Edmonton tomorrow and shall be there a day or so then to Lakefield where we shall probably be a couple of days also, then N.Y.

We have received no mail at all since your letter of May last of or from Muriel, Bryce, May, we have heard nothing since we left nearly 18 months ago and of or from Ryan I have heard nothing since I saw him off at N.Y. March 1911. It is rather curious coming back to the "civilized" world.

Melville is here and coming to Edmonton tomorrow, he may be in N.Y. in a month or so.

Love to all, hoping you are all well and looking forward to seeing you shortly.

Always yours affecty.<br>George M. Douglas.

By Telegram from Edmonton: October 18, 1912

James Douglas
99 John St,
New York City

Arrived here today all of us well Left Bear Lake with dogs April 29th arrived mouth of Coppermine on foot June 5th walking back to Bear Lake Trip successful no startling discoveries A long canoe voyage across Bear Lake account ice Wire King Edward Hotel, Edmonton write Lakefield going there direct then to New York

George M. Douglas

www.ingramcontent.com/pod-product-compliance
Lightning Source LLC
Chambersburg PA
CBHW060423100426

42812CB00030B/3290/J